CW00548995

CONTENTS

INTRODUCTION
ACKNOWLEDGEMENTS
DEDICATION

Massage has a fascinating history dating back 4000 years to China, and other great ancient civilizations such as those of Japan and India. The ancient Greeks and Romans also promoted massage both for sport and remedial health. Hippocrates of Cos (460-380 BC), who is generally accepted as the 'father of medicine' and who authored the Hippocratic oath, declared that physicians should be experienced in many things, one being "... rubbing, as this can bind a joint that is too loose and loosen a joint that is too rigid ..." This principle is still applied today within physical therapy. Hippocrates' teachings were passed on to other famous physicians and philosophers; one of whom, Galen (130-201 AD), wrote books on sports massage, and was reputedly a physician to the gladiators of the Circus Maximus in Rome.

During this era, however, massage became more notorious for its erotic benefits, causing a lowering of professional standards and resulting in some of today's misconceptions about the practice. Then, during the Dark Ages that followed the fall of Rome, religious leaders dismissed the touch-based healing methods of Hippocrates and Galen, regarding them as inappropriate and sinful.

Massage was probably 'reintroduced' by the innovative French physician, Ambroise Pare (1510-90), who used it as a valuable form of complementary medicine. In 1813, the Swede Pehr Heinrick Ling brought massage back to mainstream complementary therapy: working within the world of gymnastics, he devised what we now call 'Swedish Massage.' In the late 1890s, a professional body of masseuses was formed in Britain, and granted a Royal Charter. It later amalgamated with the Institute of Massage and Remedial Gymnastics, and, in 1944, became the Chartered Society of Physiotherapy (CSP).

Massage and your dog

Historical narratives regarding animal massage are few and far between. However, it is believed that the ancient healer Arrian recommended massage for horses and dogs, recording that

The importance of reassurance through touch is a fundamental aspect of any massage.

"... it would strengthen the limbs, render the hair soft and glossy, and cleanse the skin." Evidence of the beneficial effects of massage on animals is not difficult to find, however; we need only look at animals themselves to understand the importance they place on the use of massage through mutual (allo-grooming) or self-grooming. Among canines, an individual suffering from a sore or some other form of injury will often have the troublesome area roughly licked by another pack member.

This action, which resembles rubbing, is in effect a form of muscle massage. Sadly, when canine massage is mentioned, the image frequently conjured up in the minds of many is that of the 'pampered pooch.' Even if pampering *was* the only benefit, that's not necessarily a bad thing, but we should not allow this perception to colour the very positive and beneficial qualities of canine massage.

Massage will enable you to gain a greater awareness of your dog's overall condition.

The techniques explained and demonstrated in this book should assist you in improving your dog's health, and help you gain a greater awareness of his or her overall condition. Other techniques are also used by fully trained and qualified professional canine myotherapists or massage therapist practitioners, but the ones included here are safe to use without such specialist knowledge, if followed according to the conditions highlighted. In other words, the important massage mantra 'do no harm' applies.

Massage has many qualities and applications. For example, it can induce relaxation or promote excitability; it can aid the removal of cellular toxins as well as enhance the delivery of oxygen and nutrients to the cells. It can also profoundly relieve musculoskeletal problems, and those impacting on mobility caused by a medical condition or disease. Post-operatively, the remedial benefits that can be felt from the correct use of massage are immense, bringing about a quicker and more effective recovery. Massage is also a positive way of communicating and being more 'in touch' with your dog.

Relaxation

The relaxation benefits derived from massage are linked to the way in which the body's hormonal system functions. Applying the correct form of touch and gently manipulating your dog's muscles will immediately induce the release of hormones known as endorphins. These hormones promote a feeling of well-being and help to inhibit pain: dogs and horses seem especially receptive to them. (Massage has other calming effects that are explored in more depth later in the book.)

Excitability

Working and performance dogs, such as police dogs, sheepdogs and dogs involved in agility or obedience, have especially active lifestyles. By applying a specific type of massage we can initiate a neural (nerve) response that will alert their brain to an 'imminent event,' and also warm their muscles to be ready for action. This can help reduce the incidence of injury as well as improve the flow of blood, oxygen and nutrients to the muscles to help provide and sustain healthy activity.

Musculoskeletal problems

The muscular system is one of the ten major systems of the body; other important ones are the digestive, circulatory, skeletal, respiratory and endocrine (hormonal) systems. It is vital that all body systems operate correctly to ensure good health and, indeed, ultimately survival. These individual body systems are interdependent; a body needs each of them to be functioning correctly for the 'whole' to be working efficiently. A malfunction in one system can seriously affect the proper working of another. A common yet underestimated malfunction can be a muscular – sometimes described as a musculoskeletal – problem; in this case, a damaged muscle can have an adverse effect on the body's skeletal system, just as a problem within the circulatory or digestive systems can have a similar consequence.

When musculoskeletal problems are identified, massage can influence change within muscles very effectively through the use of different techniques, depending on the result you wish to achieve. Muscle damage arises through different situations: injury, compensatory issues, or repetitive strain. All of these separate, yet overlapping, issues can be addressed through the correct application of different massage techniques.

Massage can enhance muscular balance and health for all dogs.

Post-operative or remedial treatment

Massage has a cleansing effect that helps remove toxins from the body and promotes the same 'feel good' factor that one gets after exercise. Providing this feeling through massage in a dog who is recumbent – perhaps in a post-operative situation or because he lacks good mobility – will enhance general health, body function and metabolism, in addition to providing a relaxing experience.

Correct body function is only possible if all of the cells in the body are also working correctly or are 'mobilized.' It's apposite to think of every cell in our body as a 'mini-me': each cell needs to take in food, build itself up, lose waste products, take in oxygen, expel carbon dioxide, and reproduce. If any of these actions are impeded, the cell will not function properly, and if too many of the cells in an organ or a system are not working, the effects can be serious.

Musculoskeletal issues can occur in all sizes and types of dog. Here, a Leonberger is being given massage treatment.

Communication through touch

The need for us to bond with our dogs is usually deep and profound, and massage can provide one of the best methods for achieving this. For all of us, the intrinsic requirements for survival are essentially the same: a feeling of safety and security, plus food and comfort. However, dogs can become stressed on a daily basis in situations that would seem normal and even comforting to us!

Because most of us live in a hostile-free environment, it can be difficult to empathize with the daily need for security experienced by a puppy, or a recently re-homed dog; for them, this need is very real. Consider what is must be like to be shut in a room or a car for hours, unable to

This older dog is displaying muscular stiffness through the back legs and compensating through the front legs.

release yourself, trusting that someone will return to free you, feed you, and comfort you. This is possibly how dogs perceive the domestic situation we place them in. Given this potential for daily stress and anxiety, a firm and secure bond between dog and handler can reduce these and the other stresses of living in a human environment.

Scientists have shown that massage can prime the human brain to release oxytocin, a hormone that assists interaction with strangers and encourages bonding and trust; this may have the same effect on dogs when we massage them. This is just one way in which massage and touch can positively enhance the connection between dog and human; others are explored in the chapter entitled *How does massage work?*

Please note that, for ease of reading, this book refers to the dog as male throughout (unless a gender-specific behaviour is under discussion): however, female is also implied at all times.

Acknowledgements

I would like to thank Marc Henrie and Henry Robertson for supplying the fantastic photos for this book (unless otherwise indicated).

Heartfelt thanks, too, to all of the extremely patient dogs and their handlers, who kindly gave up their time to help.

Dedication

To all the dogs, both present and sadly passed, who I have had the pleasure to treat; I would like to thank them for all that they have shared with me and taught me

When not to massage

Always be sensitive to your dog's needs when giving any type of massage. Massage is intended to be consensual; in other words, something both you and your dog are happy about engaging in. It is important to note the circumstances and occasions when you should not try to apply massage.

- If your dog gets up and walks away
- If your dog displays any signs of pain or obvious discomfort
- If your dog is ill

Never force your dog into a situation where he feels trapped and cannot escape, because this will promote his flight instinct, and can induce defensive or aggressive behaviour.

By contravening any of these three very basic rules you can inflict harm on your dog, or create a situation that will negate any trust that has developed between you. These contraindications (situations when massage is inadvisable) are also explored more fully in the chapter entitled How does massage work?

1
HOW THE DOG WORKS

The body of a dog, like the bodies of most other animals, is a complex, living machine. And as explained in the Introduction, the body relies on all its different systems operating in harmony to ensure proper function and good health.

Massage is an effective method of helping to repair parts of the body 'machine' when it is faltering, but to maximize the benefit of the massage treatment you are applying, it's essential to understand a little of how the body of a dog works in biological terms. This can assist in not only determining the nature of a specific problem, but also – more importantly – help you understand how the massage you are applying will affect the canine patient.

A body in balance

Just like our own systems, those of your dog are finely balanced. The body constantly monitors its internal environment, making small adjustments to keep everything functioning at the optimum level, irrespective of external conditions. This series of ongoing regulatory processes that ensures the

The location of the major skeletal areas or bony landmarks of the dog. You can feel these when you examine your dog.

Heavy panting indicates a lack of equilibrium or homeostasis.

body stays in equilibrium is called homeostasis; essentially, 'keeping things balanced.' The system uses a negative feedback mechanism that is triggered whenever there is a deviation from the normal range of, for example, blood sugar levels, water content, blood pressure, oxygen levels, temperature, and so on. Any deviation initiates a response within the body to counter the change and restore balance as quickly as possible. The way it works can be likened to the thermostat on a radiator that controls and maintains the temperature within a room. For example, if a dog gets too hot, his temperature is lowered to the correct level using the body's thermoregulatory system. The dog loses heat through his tongue by panting, while blood vessels within the skin dilate to aid cooling. The homeostatic mechanism is also involved in maintaining a correct balance during times of excitement, stress or relaxation.

At the cellular level the bodies of almost all animals, including, of course, the dog's, are made up of millions and millions of tiny structures called cells. Groups of similar cells combine to form tissues like muscles and bones, and different tissues combine to form organs like the heart and lungs. Ultimately, therefore, all of the systems in the body operate at the cellular level. Each cell must be in balance, with the correct level of salts, oxygen, minerals, and so on. When we massage, we

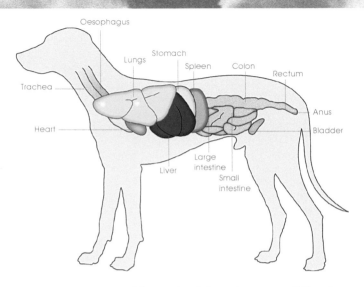

The major internal organs of the dog.

All the systems of the body have a direct influence on the muscular system, as shown by this diagram. By influencing the muscular system through the skin, we can have a positive effect on other body systems.

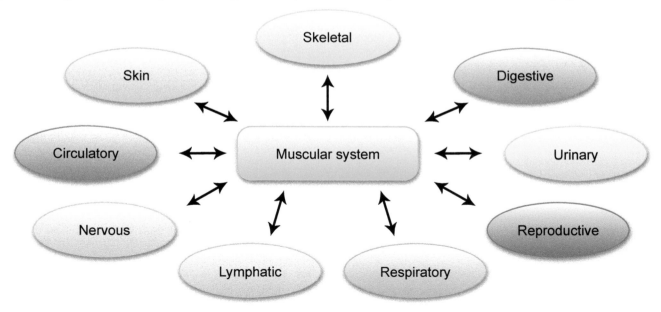

influence the body on a cellular basis by changing the environment within and surrounding different cells to produce a synergistic effect on the whole area. By using massage, we can have an effect on blood cells, muscle cells, nerve cells, and the cells that comprise the skin. Massage can even have a positive effect on the cellular development of the skeletal system. Let's take a look at each of the major systems within the body.

The skin

The skin forms the outer body covering of the dog, and is one of the body's major organ systems; in fact, the biggest organ in the body. As befits such an important structure, the skin has several different roles. Firstly, it acts as a waterproof, physical and chemical barrier, helping to prevent harmful germs or other substances entering the body.

The skin also has a protective role in the form of pigmentations that help provide protection from solar radiation. Blood vessels in the skin help regulate heat: hair follicles in the skin also assist in the regulation of body temperature; the hairs can be raised or lowered, altering the amount of air trapped near the skin.

In the dog, the hairs are also raised or lowered as social signals to other dogs. Scent glands in the skin are used for marking territory, and other glands are responsible for attracting the opposite sex for mating. Whiskers around the mouth help the dog to sense the environment. Other neural (nerve) receptors in the skin provide further information about the environment, such as the external temperature. They also sense pain, helping the dog to avoid serious injury.

The skin is vital for the production of Vitamin D, and is also involved in antigen stimulation.

The skin and massage

We use the dog's skin as a conduit or connection when we massage. The skin has specialized neural

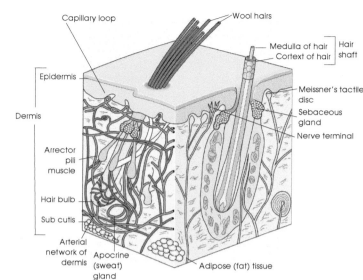

A cross-section of canine skin.

receptors, via which we can elicit changes to the body both mechanically and reflexively through touch (see the chapter entitled *How to massage your dog*). These neural receptors connect with the matrix of nerve fibres that run throughout the entire body, and can detect movement, touch, temperature, pressure and stretch signals given through the hands when massaging. Within the superficial fascia, the tissue layer just beneath the skin, lie additional neural receptors that also connect throughout the entire body of the dog. Thus, with this matrix of neurological connections, the effects of massage can be transmitted to almost any part of the body via the skin. The action of massage can also aid natural drainage of the sebaceous glands, helping to remove toxins and keeping the coat healthy.

The circulatory system

The body's blood circulatory system can be

divided into two sections: the arterial section and the venous section. The arterial section is the delivery system: most arteries carry oxygen-rich blood to the cells. The blood is pumped by the heart through an extensive network of arteries; these are wide, muscular vessels (tubes) that pulse the fluid to all parts of the body under high pressure. The arteries then divide into thinner vessels called arterioles, and the blood – now under less pressure – is delivered to the cells of the body. At this point the blood passes into the extremely small capillaries that form a capillary bed surrounding

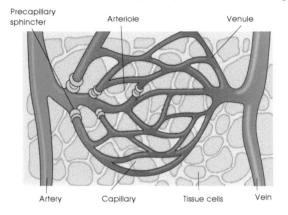

A small capillary bed, demonstrating the transfer of blood from the arterial delivery system to the venous waste disposal system.

the organs and tissues, from where it passes directly into the cells by osmosis. While the arterial blood is being delivered, the reverse procedure is also taking place, and venous blood is taken from the cells.

The venous section of the circulatory system is the body's waste disposal department, removing the toxic products of cell metabolism, such as carbon dioxide (metabolism means the physical and chemical processes necessary for life). The venous system transports the blood from tiny capillaries into progressively larger vessels – the opposite to the arterial system. Instead of using the heart to provide high pressure to move the blood

through the veins, the system relies instead on skeletal movements, movements of the diaphragm causing internal pressure changes, and the effect of the pumping action of the main arteries which are in very close proximity to some of the main veins.

The venous return has a more difficult and no less important role than the arterial delivery, since it must carry the body's toxins through a complicated network of vessels as quickly as possible to organs such as the liver and kidneys, and then to the heart and lungs, where they can be rendered harmless or expelled from the body before the blood is reoxygenated by the lungs and pumped back through the arterial system.

The venous system relies on several main methods for transporting the blood:

1   Skeletal movement
2   The position of larger vessels near to main arteries means that their pumping action has a small knock-on effect
3   Internal pressure changes (diaphragm movement)

How massage influences the circulatory system

The reason why massage is so beneficial in aiding blood circulation is found by examining the processes which influence the venous return. Of these, skeletal movement is the most important and efficient way or enhancing the process. Therefore, when we massage the superficial muscles, we are gently replicating skeletal movement. By aiding the removal of toxic venous blood within your dog's system, you will facilitate nourishing arterial delivery instead. Furthermore, venous blood can collect within the body, especially if there is a lack of activity or function, and this in turn will suppress healing. The introduction of arterial blood will have a positive effect on the wellbeing of the body. Also, by encouraging relaxation and facilitating deeper breathing, massage helps to increase the fluctuation of internal pressure changes, and enhance flow through the more relaxed muscles.

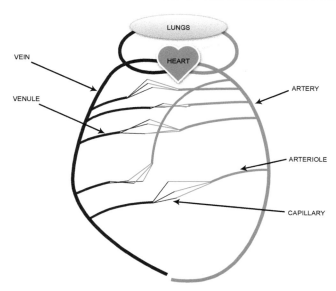

A schematic plan of the dog's blood circulatory system.

The nervous system

The nervous system passes messages from one part of the body to another. Masterminded by the brain, the nervous system controls, coordinates and directs activities such as perception, thought and movement. Some of the messages may carry sensory (feel) information to the brain, and others may transfer motor (movement) information from the brain to the muscles. In this constant and busy 'motorway' of messages, there will be instructions for both voluntary actions and involuntary reactions.

The nervous system has two main parts: the central nervous system and the peripheral nervous system. The central nervous system consists of the brain and the main nerves travelling down the spinal cord. These nerves receive messages (information) from the body, which are passed to the brain. The brain assimilates the information and transmits instructions to the nerves of the spinal cord, which distribute the instructions to other parts of the body.

THE NERVOUS SYSTEM	
Central nervous system	Consists of the brain (equivalent to the hard drive on a computer, it retains the information and programmes), and the main nerves travelling down the spinal cord. These nerves distribute messages from the brain and receive messages and information from the body. The brain assimilates and reacts
Peripheral nervous system	Consists of the nerves that arise from the vertebrae and serve the body's systems and organs (equivalent to the software of a computer, it performs the actions). Within this system lie three different divisions:
Sensory	Inner body sensations sent to the brain: pressure, heat, cold, movement, position
Motor messages	Sent from the brain to initiate voluntary movement, to adjust to situations and surroundings depending on stimuli. Instruct voluntary muscles to contract and relax
Autonomic (involuntary)	Sympathetic (fight or flight) or parasympathetic (rest and digest). Controls internal organs and blood vessels

The peripheral nervous system consists of the nerves that arise from the vertebrae and serve the body's organs and tissues. Within this system lie three different divisions. Between them, they are responsible for providing sensory information experienced by the body (such as pressure, heat, cold, movement and position), for the contraction and relaxation of the main muscles to produce movement or to respond to environmental conditions, and for the involuntary movements of the body in response to danger, as well as for actions such as the rhythmic beating of the heart.

From the spinal cord a pair of spinal nerves leave the main 'highway' of the central nervous system and innervate structures such as the forelimbs and hind limbs.

The spinal nerve is itself divided into the dorsal (or upper) roots, and the ventral (or lower) roots. The dorsal root carries sensory fibres towards the spinal cord, passing information from the body's extremities to the brain for assimilation. The ventral root carries motor fibres to the limbs to initiate movement. It is not possible to repair nerves within the central nervous system, so any damage here can be catastrophic in its effect, whereas injury

Dogs at rest, displaying the relaxed state of 'rest and digest' influenced by the parasympathetic nervous system.

Dogs active and alert, displaying the excitable state of 'fight or flight' influenced by the sympathetic nervous system.

The important cranial nerves supply and receive both sensory and motor impulses simultaneously. They are still part of the peripheral nervous system but are named according to their function. They are numbered according to the order in which they arise:

THE CRANIAL NERVES

1 Olfactory* (smell)
2 Optic (sight)
3 Oculomotor (eye movement)
4 Trochlear (fine eye movement)
5 Trigeminal (sensory to forehead, nasal cavity and cornea; motor to jaw)
6 Abducens (motor to eye)
7 Facial (facial expression, taste, salivation, tear production)
8 Vestibulotrochlea (hearing and balance)
9 Glossopharyngeal (taste, swallowing, salivation)
10 Vagus* (swallowing, vocalization; also parasympathetic fibres to heart, lungs, liver; slows heart rate and allows digestion)
11 Accessory (movement of neck, swallowing, vocalization)
12 Hypoglossal (movement of tongue and swallowing)

Areas believed to be influenced by massage

to the peripheral nerves is more likely to show as a progressive lack of function within muscles, as well as loss of tone.

The type of stimuli detected by sensory receptors in skin, muscle and tendons include: pain, temperature, stretching, touch (pressure, vibration and texture), and proprioception. (Proprioception is a kind of additional sense that provides the body with a perception of its position, movement and posture in physical space.) The importance of this mechanism is that it is a protection device, which affects, in one respect, the position of muscles and joints, the mechanism which keeps the dog standing upright and responding to both internal and external environments. It is further supported by neural mechanisms within the joints, tendons and muscles which prevent overstretching – either within an injury situation or in everyday movement as a joint protection device.

take place without the need for conscious control, such as the activity of smooth (involuntary) and cardiac (heart) muscle. It also regulates heart rate, increases glandular secretions, controls the digestive process, and erects and flattens the hairs on the body.

The autonomic nervous system is divided into the parasympathetic nervous system and the sympathetic nervous system. The two systems are known more commonly as fight or flight and rest and digest – depending upon the activities with which each system is concerned – and either is initiated when the dog responds to a particular situation.

The sympathetic nervous system operates when the dog is in a proactive state ready to fight or run from danger; the parasympathetic system is initiated when the dog considers its situation to be safe, relaxed and threat-free.

The autonomic nervous system
This system is extremely important because it controls all of the activities within the body that

How massage influences the nervous system
Massage can influence various parts of the nervous system – mostly reflexively – by calming or reducing

SYMPATHETIC NERVOUS SYSTEM Fight or flight	PARASYMPATHETIC NERVOUS SYSTEM Rest and digest
Eyes focus on distant objects Trachea opens and blood vessels of lungs dilate Heart rate increases Liver releases glucose Reduction of digestive enzymes Digestive system slows down – blood focused away from this system Bladder sphincter constricts	Eyes focus on near objects Trachea and bronchial tubes constrict Heart rate drops Liver stores glucose Secretion of insulin Digestive system speeds up – blood focused on this system Bladder sphincter relaxes

The two parts of the nervous system work in opposing ways to protect or facilitate appropriate changes to ensure the dog's survival.

pain perception, thereby easing stress and thus promoting a feeling of calmness and security. It will therefore stimulate the parasympathetic system, facilitating enhanced digestive processes.

Through easing muscle tension, the stretch and pain receptors within muscles can also be eased, producing great integrity within the muscle and also the joint action, improving 'range of movement.'

The use of passive movement and other 'mechanical' techniques will help re-educate the stretch receptors within the muscles and joints to allow greater movement. This can assist performance, and is especially beneficial in remedial situations – for example, in post-operative and post-treatment – and in the chronically lame dog.

Massage can assist in the facilitation of a full range of movement, so that the dog's natural activities are not restricted.

The muscular system

Muscles form the bulk of the soft tissue in the body. The complex muscular system has two components: the involuntary muscles, mentioned earlier, encase organs such as the digestive system, the arteries, heart, and other internal organs. The dog cannot directly influence these muscles through conscious thought. The skeletal, or voluntary, muscles are attached to the skeleton, and provide the forces necessary to move the limbs through an ingenious system of levers, enabling the dog to walk, run, move its head, and so on. The skeletal muscles also help provide a sound foundation for the joints so that the dog can perform changes of movement and direction without causing itself injury. The skeletal muscles also help maintain good posture, so that, even when resting, the body can support itself and function efficiently. It is vital that all of the skeletal muscles work in harmony if the muscular system is to operate effectively as a whole. Furthermore, the muscles help to generate heat when the animal shivers. The involuntary muscles also control heat regulation by vasodilatation (enlarging the blood vessels), and by vasoconstriction (narrowing the blood vessels), as well as by using the tiny muscles within the skin called arrector pili muscles – see illustration of skin section, page 13 – to lift the coat to trap air and provide extra insulation.

For the body to move effectively, all of the

A good example of a dog displaying muscular balance and stability.

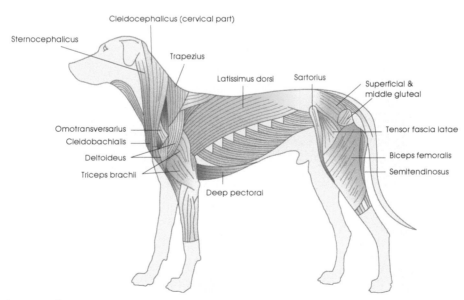

Some of the major muscle groups in the dog.

skeletal muscles must be balanced: in other words, the muscles on the left and right sides of the dog should be of equal length and working to full efficiency. Furthermore, it is important that 'phasing' – the sequence of contraction and relaxation of the muscles – and the correct

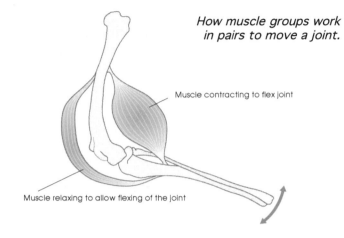

How muscle groups work in pairs to move a joint.

Muscle contracting to flex joint

Muscle relaxing to allow flexing of the joint

balance of stability provided through the joints is equal to the movement required. This will ensure that the dog is using the back legs, pelvis and

lumbar region to drive the body forward, while the front legs and pectoral region absorb concussion and control direction.

The skeletal muscles can be divided into two further groups: deep muscles and superficial muscles. The deep muscles, generally lying closest to the joints, tend to have a supporting role. The superficial muscles are attached to the joints and provide the power for movement.

Muscle structure and function

All muscles comprise single, elongated cells, known as fibres, grouped together in bundles. These are further grouped together in larger bundles that contain whole muscles, which, in turn, are linked with other muscles to form muscle groups. The muscles are attached to the bone by means of tissue known as tendons, or by specialized connective tissue. The action of moving a limb – for example, lifting a leg – involves groups of individual muscles working together in opposite pairs, with the actual movement taking place at a joint. First, one set of muscles (the agonists) contracts, flexing or bending the joint. The muscles lying opposite

Displaying muscles that are relaxing to achieve maximum movement range.

Displaying muscles that are in full contraction to achieve maximum power.

the joint (the antagonists) simultaneously relax to allow the flexion. When the joint needs to extend in the other direction, the opposite happens: the original antagonist muscles now become the agonists since they are initiating movement, and the agonists become the antagonists and relax to allow the extension. When the muscle groups work smoothly to provide movement, this is called coordination.

Muscle contraction

All muscle fibres are made up of millions of small filaments, of which there are two sorts: thick filaments made from the protein myosin, and thin ones made from the protein actin. The muscle contraction process begins with a signal from the brain that goes from the nerve to the muscle. The nerve endings release a chemical that causes the actin filaments to slide between the myosin filaments, pulling the ends of the muscle towards each other, and thus shortening it.

Muscular tension

Muscular tension can occur through an injury or some other condition. Tension generally causes pain and a subsequent imbalance in the body's musculature, and may bring on conditions such as lameness. The inhibited action of the tense, sore muscles puts stress on joints and also other muscles; this, in turn, can cause further pain and even more stress.

The ensuing damage to the affected muscle fibres will compromise the mineral balance within the fibre that is vital for healthy contraction and relaxation of the muscles, which can lead to problems with muscle tone and function.

(In this case, muscle tone refers to the amount of individual fibres functioning within a single or group of muscles; the fewer fibres able to function, the lesser the strength or functionality of the muscle.) Contributing to this pain perception and

Repeated exercises, including chasing games, can be a major cause of repetitive stress, and can give rise to compensatory issues.

Jumping down from chairs and cars can have a negative effect on the dog's muscular balance, and therefore its mobility.

lack of function through tension and imbalance will possibly be subsequent inflammation due to additional joint and muscle stresses. This will contribute to further injury or damage to existing or new areas through a continuation of this process. The chain of events following an injury or other causal condition is often referred to as 'compensatory issues.' These clearly have a negative effect on the dog's muscular balance, and therefore its mobility; in turn possibly causing further lameness and also impacting on the body's homeostasis.

How massage influences the muscular system

The superficial muscles are the easiest to influence when massaging. As a way of enhancing mobility, massage can be very effective in several ways, firstly, by enhancing venous return, which assists toxin removal and therefore enhances the cellular environment to encourage healing. It boosts arterial flow to provide nourishment for the muscle cells to assist healing and aid function, and the lymphatic flow to support the immune system and aid detoxification. It reduces spasm and engorgement, and therefore eases and enhances circulation and neural function. It aids relaxation and eases stress through reducing muscular tension. Finally, it replicates muscle movement, and therefore influences circulation, lymphatic flow and neural activity by helping to maintain muscle tone.

The skeletal system

The skeleton is the rigid framework of bones inside the dog's body, which provides the basis of attachment for the muscles used to move the body, as well as protecting delicate organs such as the brain, heart and liver. The skeleton is divided into two main parts: the axial skeleton and the appendicular skeleton. The axial skeleton comprises the bones of

the skull, the hyoid apparatus (larynx), the vertebral column and the ribcage – mainly the bones that provide protection. The appendicular skeleton comprises the bones of the limbs, as well as the bones that connect the appendicular skeleton with the axial skeleton – for example, those of the pelvis and the scapula.

Each bone is different, with its own distinct shape and size. Many are characterized by grooves, protuberances and holes, and both smooth and rough areas. These serve mainly as attachment points for muscle ligaments and entry and exit points for blood vessels. Parts of some of the bones are visible just under the skin of the dog, and these are known as anatomical landmarks. These bones – such as the shoulder blade or scapula – are usually attached to large muscles that provide stability or movement.

Bones begin developing and remoulding from the time the puppy is in the womb. The bones will continue to remould and change throughout the dog's life, depending on the stresses placed on, or through, them; these stresses can have a positive

The dog's skeletal system.

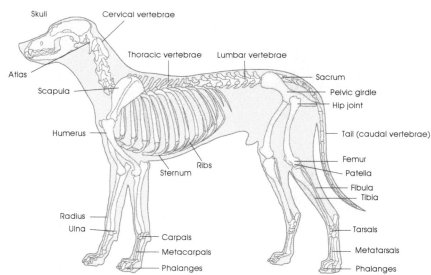

Skull · Cervical vertebrae · Thoracic vertebrae · Lumbar vertebrae · Sacrum · Pelvic girdle · Hip joint · Atlas · Scapula · Tail (caudal vertebrae) · Humerus · Femur · Patella · Fibula · Tibia · Ribs · Sternum · Radius · Ulna · Carpals · Tarsals · Metacarpals · Metatarsals · Phalanges · Phalanges

effect by developing strength and good rigidity through the bones. However, if inappropriate stresses are applied, the skeleton will not have the same robust nature.

Joints

Joints form the junction between two or more bones. Most joints provide movement and flexibility to the skeleton, although the amount of movement provided by a joint varies; in fact, some joints – those of the skull, for example – allow no, or only barely perceivable, movement. Bones such as the elbow and knee articulate at movable joints called synovial joints. These are the most common types of joint in the body, and also the most important ones for assisting and influencing healing

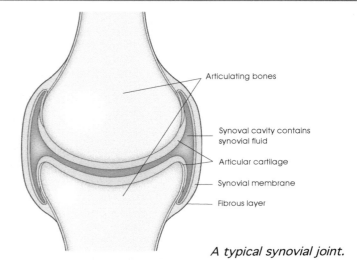

Articulating bones

Synoval cavity contains synovial fluid

Articular cartilage

Synovial membrane

Fibrous layer

A typical synovial joint.

The chart shows the most relevant body systems that serve the muscular system and can be positively influenced by massage. Optimum function shows how all the body systems work together when the muscular system is uncompromised. Reduced function shows the effects on the different systems when there are problems within the muscles – either individually or collectively.

	MUSCULAR SYSTEM: OPTIMUM FUNCTION
Nervous system	Good neural reception is required for muscles to contract and relax effectively
Skeletal system	Secure joints and muscle attachments are required for muscular, and therefore skeletal, stability
Circulatory system	Good circulation to the muscle fibres is necessary to keep them nourished and cleansed
Skin	Provides protection and easy of movement through lack of stress and tension
Lymphatic system	Secondary aid to the removal of toxins from the muscles and surrounding structures

Nervous system

Skeletal system

Circulatory system

Skin

Lymphatic system

MUSCULAR SYSTEM: REDUCED FUNCTION

Without good neural messages serving muscle fibres, the muscles will not contract and relax at their optimum, or maintain good tone
Possible causes: Disease. Injury. Compensation issues
Issues affecting the muscular system: Lameness. Stiffness. Lack of full movement. Lack of good coordination. Lack of spatial awareness. Lack of good muscular development
Issues affecting the nervous system: Pain. Stress

Insecurity within the skeletal system can cause imbalance within the muscular system, and vice versa
Possible causes: Disease. Injury. Compensatory issues
Issues affecting the muscular system: Lameness. Lack of balance. Lack of core strength Lack of even muscle tone development
Issues affecting the skeletal system: Stresses through joints, with ensuing wear and tear leading to pain

Lack of good circulation will inhibit both the delivery of nutrients and oxygen and also greatly reduce cellular cleansing, leading to inhibited growth and development
Possible causes: Disease. Injury. Compensatory issues
Issues affecting the muscular system: Reduced activity. Reduced cleansing of muscles, therefore possible damage. Premature fatiguing. Lack of balance. Lack of even muscle tone development
Issues affecting the circulatory system: Pooling of venous blood, therefore a reduced cleansing function and an inhibition of good circulation leading to pain

Tension caused through lack of good movement will reduce movement further, causing possible adhesions
Possible causes: Disease. Injury. Compensatory issues
Issues affecting the muscular system: Lack of mobility. Uneven gait. Lameness. Reduced range of movement
Issues affecting the skin: Lack of good nourishment through tension causing possible skin conditions. Lack of good condition within the dog's coat. Itchiness over affected areas leading to pain

Lack of good cleansing can cause tension within the muscles and therefore reduced function. Underactivity can cause engorgement within muscles
Possible causes: Disease. Injury. Compensatory issues
Issues affecting the muscular system: Lack of cleansing can cause engorgement that will lead to further damage of the muscle fibres, leading to further lack of mobility and pain
Issues affecting the lymphatic system: Reduced movement of the lymphatic system can compromise general health and body immunity leading to pain

through massage. Within a synovial joint, the two articulating bones are divided by a fluid-filled structure known as the synovial, or joint, cavity. The synovial fluid is contained within the joint capsule, which is composed of an inner synovial membrane and an outer fibrous layer. The head of each articulating bone is capped with a substance called articular cartilage: its purpose is to cushion and protect the bones. It does not possess its own nutrient source but relies on support from the synovial fluid. Therefore, if the synovial membrane is damaged and the synovial fluid compromised, the cartilage will also suffer and degenerate. Synovial joints are the ones that can cause arthritic problems due to friction and lack of integrity within the synovial fluid, the cartilage or the joint capsule.

How massage influences the skeletal system

By correctly applying passive movement, massage can enhance mobility through the joint and also assist joint lubrication. By reducing muscle tension around the joints, massage can ease frictional stress and pain perception, especially with conditions such as osteoarthritis, and improve range of movement, mobility and speed.

The lymphatic system

The lymphatic system is often known as the immune system. As the name suggests, its function is to protect the body against infection and disease. Lymph fluid circulates the lymphatic system and contains a high concentration of white blood cells, whose job it is to engulf potentially harmful foreign bodies. Unlike the blood circulatory system, fluid in the lymphatic system flows in only one direction – that is, from the most distal areas, such as the feet, towards the cardiac region where it empties through two main ducts into the venous system.

It also flows around the cells of organs, where it is known as interstitial fluid. In a similar way to the blood circulatory system, the lymphatic system comprises a vast network of thin vessels and capillaries that transport the lymph fluid to special sites within the body called lymph nodes. These nodes enlarge when they fight infection.

Once the lymph fluid has collected the foreign bodies, it enters the lymphatic vessels and is transported to the lymph nodes where, hopefully, the organisms will be destroyed. The lymphatic system is a vital secondary waste-disposal system that acts as backup to the venous system. The lymphatic system relies on the same processes to take the contaminated lymph fluid from the vessels in the body to the point of cleansing, therefore skeletal movement and internal pressure changes play a major part in this process.

How massage influences the lymphatic system

Massage can influence the return flow of the lymphatic system in very much the same way as it does the venous system (see page 14, *How massage influences the venous system*).

The difference is that lymphatic massage need only be really superficial, or gentle, in other words. The lymphatic system can be seen to be malfunctioning when certain areas, especially feet and legs, swell due to a condition or lack of mobility. Applying a gentle stroking massage technique or passive movement will help promote the return flow of lymph fluid), and aid the passage of interstitial fluid into the lymph vessels to ease the swelling. (If your dog has any form of abnormal swelling, you must seek veterinary advice before you attempt any type of massage.)

Massage will benefit the healthy dog, or one who is recovering from a condition, by stimulating the lymphatic/immune system.

2

USING YOUR VET

Before beginning any type of massage treatment on your dog you should consult your vet in case there is an underlying health issue that requires specialist medical attention. It's very helpful to vets if owners or handlers are observant about their dog's health, and even more so if they can produce a short, relevant history of the problem to discuss at a consultation. A diary of observations, such as abnormal heat discernible around muscles or joints, along with other relevant issues, such as lameness (both persistent and intermittent), can aid a speedier diagnosis and therefore resolution to the problem.

Should your dog ever need emergency treatment, any useful information you can provide to the vet in advance of your arrival at the surgery with your dog, including such symptoms as abnormally high heart rate or breathing rate, can again help to reach a quicker diagnosis of a problem. Such information could also mean a much quicker recovery time for your dog. You can

Your vet should always check your dog first to ensure massage is appropriate.

learn these sorts of observational skills by attending a good canine first aid course. The same is true if your dog has an ongoing diagnosed condition affecting mobility, or an undiagnosed lameness; to have 'hands-on' history can be extremely useful when trying to isolate the primary or compensatory effects, and reach a diagnosis.

It is helpful to all pet health professionals if you have some working knowledge of how your dog functions, from both a physiological (how the body works) and psychological (how the brain works) point of view. Having such knowledge means that you will better understand the advice and other information that the vet gives you concerning the problem. It also means you will have a better idea of the type of information you need to give your vet. However, no matter how much we feel we know about our dog, what we do not know will always be far greater – and that is where our veterinary surgeon is critical. Sometimes we may think that a lump, swelling, or even lameness is minor and insignificant, and that merely to seek therapeutic help is the best course of action. However, this protocol is not only misguided, it could also mean that something seemingly inconsequential could, in fact, be the sign of a potentially far more serious condition. Early detection by an expert is critical, and it must be stressed again that your first appointment should be with the vet!

Getting consent from your vet

It is especially important to provide your vet with a good condensed history of the problem affecting your dog when consulting him or her about the possibility of applying massage, either yourself or via referral to a canine myotherapist or massage practitioner. As explained in other chapters, the effects of massage are far-reaching, with the reflexive consequences often impacting on more than one of the body's systems, so it is vital that you inform your vet of all your observations before you begin any massage. It may well be the case that

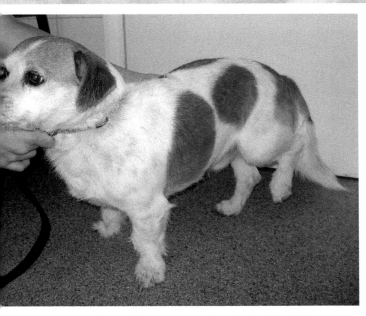

Lameness that induces a complete lack of weight-bearing ability, as here, should always be investigated as soon as possible by your vet.

Implications of the Veterinary Act (1966)

In the UK, veterinary medicine, healthcare and treatment operate in a different way to that of human medicine, healthcare and treatment. This the direct result of a parliamentary act passed to protect the welfare of sick and injured animals, including, of course, dogs. Within human medicine, it is not a legal requirement that people seek their GP's referral before having complementary therapies, even though it is recommended. The situation is different, however, with regard to animals.

The work of British veterinary surgeons is governed by the Veterinary Act (1966), and under the act no one (with certain exceptions) may practice veterinary surgery unless they are registered with The Royal College of Veterinary Surgeons. (Here, the definition of 'veterinary surgery' also covers areas of physical manipulation that include such therapies as massage, physiotherapy and chiropractic osteopathy.) The only exceptions to this ruling are when the treatment is carried out by the owner of an animal, by other members of the owner's household,

Running in a well-matched pack is a natural activity for a dog.

sometimes massage can have a negative, rather than a healing, effect: for example, massage can be detrimental if your dog has been injured in some way, especially if the treatment is applied incorrectly or at the wrong time. If your dog has suffered a physical injury and is demonstrating pain by limping or a change in behaviour, you should always seek veterinary attention in the first instance. Sometimes in these situations owners decide not to go to the vet since they feel that being prescribed some form of medication will not have a positive effect on the overall recovery and resumption of mobility. However, the injury can sometimes be such that, without medical intervention, any inflammation may be prolonged, elongating recovery and resulting in an ongoing problem. Therefore, a vet should initially check all injuries, to ensure that the intended healing remedy is credible and progressive.

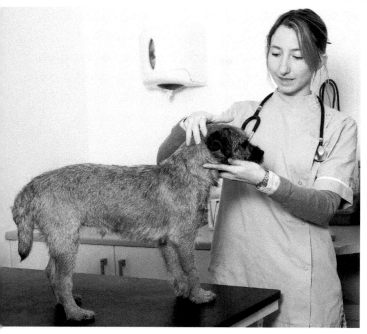

No matter how much you feel you know about your dog, always be guided by the knowledge and experience of a good vet.

or when lay persons administer first aid in an emergency for the purposes of saving life or relieving pain or suffering.

As a result of The Veterinary Act (1966) and the subsequent exemptions, it is illegal for any person, other than the owner of the animal, to treat an animal unless the permission of the animal's veterinary surgeon is sought and obtained.

Unfortunately, although this law is often not understood or upheld by some therapists who practise canine therapy, it is nevertheless there to ensure 'best treatment' for all animals.

What does a 'duty of care' mean?

All who have animals in their care – including dog owners or handlers – must, by law, provide a 'duty

of care.' Before the Animal Welfare Act (2006), people with animals only had a duty to prevent suffering, but now there is a more prescriptive definition of this obligation. This relatively new law could potentially have implications for any handler who flouts the Veterinary Act (1966) by seeking treatment from any therapist before receiving legal consent from the dog's acting vet.

The Animal Welfare Act specifies five areas of pastoral care that must be given to all animals. These are called the Five Freedoms:

- Freedom from hunger and thirst
- Freedom from discomfort
- Freedom from pain, injury and disease
- Freedom to express natural behaviour
- Freedom from fear and distress

The canine team

The best type of healthcare for your dog is one that

It is illegal for anyone to treat your dog without veterinary consent.

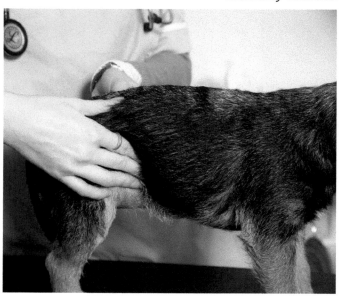

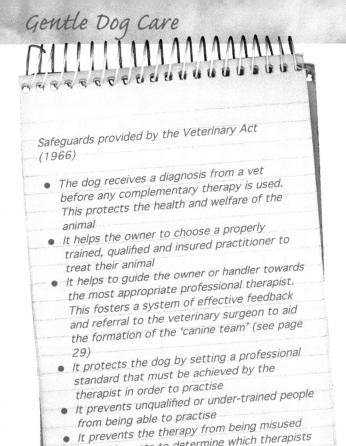

Safeguards provided by the Veterinary Act (1966)

- The dog receives a diagnosis from a vet before any complementary therapy is used. This protects the health and welfare of the animal
- It helps the owner to choose a properly trained, qualified and insured practitioner to treat their animal
- It helps to guide the owner or handler towards the most appropriate professional therapist. This fosters a system of effective feedback and referral to the veterinary surgeon to aid the formation of the 'canine team' (see page 29)
- It protects the dog by setting a professional standard that must be achieved by the therapist in order to practise
- It prevents unqualified or under-trained people from being able to practise
- It prevents the therapy from being misused
- It enables vets to determine which therapists are most appropriate to use for the conditions presented

is professional, cohesive and extensive. The most effective way to achieve this is to ensure that any referrals for your dog remain in contact with your vet, informing them of treatment and any progress. This ensures that your dog receives 'joined-up' care and that everyone has full knowledge of all treatments given.

It's also important that the therapist chosen to undertake treatment informs you fully about the treatment protocols being used, as well as suggesting others that may be available to reduce pain and help treat the condition. This could, for example, mean suggesting another type of therapy to work alongside, or instead of, the one the dog is already receiving. Any suggestions would be subject to agreement with your vet, and subsequent reports being sent to relevant parties involved with the treatment thus far. By ensuring this type of professional standard, it's possible to develop a significant and effective 'canine team,' all communicating and working in the best interests of your dog.

3

HOW MASSAGE WORKS

Massage has become generally accepted, and its benefits more and more appreciated. Treatment carried out on many dogs over the years has shown real and evident benefit, further substantiated by the fact that, since – as far as we know – dogs do not possess the ability to distinguish the placebo effect (in other words, the 'expectancy' that a treatment will work), it's reasonable to assume that the changes – both physiological and psychological – seen within the canine as a result of massage are actual.

Massage is known as an holistic treatment, because it has a comprehensive affect on the whole body. Touch is a vital part of the everyday lives of many animals, including dogs and humans; it conveys non-verbal messages are critical for its development. In domestic situations, the puppy is removed from its 'pack' at eight weeks, which is much younger than would be the case in the wild. Consequently, such dogs come to depend entirely on humans for their safety and nourishment.

A major part of their security is the fundamental beginning of how massage works.

A distinguished veterinary surgeon and animal psychologist once observed: "Touch, by someone to whom they are emotionally attached, can be as rewarding to a dog as being given a morsel of food."

Touch produces an autonomic or involuntary response; a soft, gentle touch creating a relaxed state of mind and body. When, muscles relax, blood flows more freely through them, aiding the delivery of oxygen and nutrients. As a result, digestion is also aided by the stimulation and production of digestive enzymes and insulin. In a relaxed state, the body will also release other life-preserving hormones that a stressed body will not facilitate.

Another excellent example of how massage works is the way we respond to pain and injury management. The instinctive reaction to pain caused through an injury is to rub the affected area, an action which helps to soothe the sensory nerves which relay pain messages to our brain, and also helps to disperse subsequent inflammation. As mentioned previously, dogs react in a similar way to pain or injury by licking the affected area on each other or themselves.

The holistic effects of massage

This concept is often difficult to comprehend, but it is vital for anyone using massage to understand. The table (opposite and overleaf) shows the holistic effect of massage.

The effects of massage on the main systems of the body

The influence of massage on the circulatory and lymphatic systems is key to how massage can produce positive change. As described in the chapter entitled How the dog works, both the lymphatic and venous return systems rely heavily on skeletal movement to allow effective cleansing of the body. With the use of appropriate massage techniques, skeletal muscle movement can be replicated to stimulate both these systems,

A gentle, reassuring stroke is the fundamental beginning of massage.

	THE HOLISTIC EFFECTS OF MASSAGE
Skeletal system	Massage has an indirect effect. Muscles will be improved; therefore, there will be less stress on the skeleton, particularly around the joints. Better muscle tone improves skeletal posture. Passive movements mobilize joints. By aiding circulation, nutrient delivery to bones and joints is enhanced, enabling more efficient cell regeneration or development
Muscular system	Enhances venous return, therefore the removal of metabolic waste and toxins, and thus the reduction of stiffness within muscles or muscle groups. Aids recovery through enhanced arterial delivery, providing improved oxygen and nutrient delivery. Physical manipulation of muscles helps flexibility by easing inappropriate and damaged fibre configuration. By easing muscle tension, spasm and adhesion, massage assists and promotes full muscle function (both contraction and relaxation), therefore improving muscle tone and posture
Circulatory system	Enhances circulation by assisting venous return through replicating muscle movement that will create an enhanced opportunity for arterial blood flow to the area. Assists modulation of heart by easing stress. Improves delivery of nutrients and removal of waste products. Appropriate hyperaemia (increased blood) assists targeted muscle healing and change. Enhanced circulation can result in better hormone delivery
Nervous system	Massage can either be invigorating or relaxing. A stimulating massage will awaken the nervous system, energizing the body. A relaxing massage will calm the nervous system, thus removing the effects of stress. Massage can help the restoration of balance between the sympathetic (fight/flight) and parasympathetic (rest/digestion) nervous systems. By easing muscle tension and producing stimulation to specific areas, massage will aid neural pathway development and/or repair. By using passive movement techniques, neural 'muscle memory' and thus the range of movements in joints can be improved
Digestive system	Massage enhances circulation, and therefore better delivery of nutrients to cells around the body. It aids the elimination of waste. By easing stress, it encourages parasympathetic nervous system, therefore promoting digestion
Urinary system	Encourages more efficient waste removal, therefore encourages urine production and elimination
Respiratory system	Massage promotes relaxation in the body, which results in deeper, fuller breaths. Therefore better and more efficient gaseous exchange is promoted, and more oxygen can be delivered to cells

Lymphatic system	Massage helps the physical movement of lymph through the lymphatic system. It helps with the removal of waste, thereby supporting the immune system
Endocrine system	Massage results in less stress, leading to fewer stress hormones. It encourages the release or endorphins and enkaphalins – the body's natural pain-easing hormones. The delivery of hormones is improved by enhanced circulation
Skin	Tension through the skin and fascia tissue below can be eased with regained elasticity, resulting in better blood flow (demonstrated by enhanced coat condition and lie)
Reproductive system	Removal of stress will enhance production of reproductive hormones and aid conception

improving their capacity. This is a valuable, highly-effective and gentle enhancement of these systems which requires only moderate manipulation of the superficial muscles. By applying techniques involving slight pressure, such as effleurage, passive movement and gentle kneading, the efficiency of the venous return and lymphatic systems will be further enhanced. Via the venous system, toxins will be more quickly delivered to the liver where they will be removed, while the lymphatic system facilitates bacteria and toxin suppression and destruction of other harmful bodies.

Another way to explain how massage influences the different body systems is by separating it into reflexive and mechanical actions. Unfortunately, it is not possible to classify techniques in such a way, since many have both reflexive and mechanical effects.

Reflexive actions

Reflexive actions work indirectly on the nervous system by triggering the sensory nerves. The first sensory nerves to be affected are those of the skin, which respond to touch, temperature,

movement, and pressure. When massaging, your hand forms the first connection through the dog's skin via the coat. There will be an instant response from the dog as a result of the heat and pressure or movement of your hand. All of these sensory receptors will pass information to the dog's brain.

For example, the passive touch technique, applied by placing your hand on a specific area of the dog, will have an immediate connection through touch interpreted as pressure. The type of touch or pressure that is exerted will immediately stimulate a response, depending on where you have placed your hand and the pressure used. It could therefore result in a calming, excitable or concerned reaction in the dog, each of which will elicit a different physiological and psychological response in the animal. This is known as a reflexive response.

The warmth of your hand causes another reflexive response; the heat will influence the superficial blood capillaries within the region, causing a mild dilation or expansion of the vessels, and allowing more blood to flow; this will elicit a slight increase of blood to the area, producing a healing effect.

The reflexive effects of massage can promote release of the body's endorphins.

Mechanical actions

Mechanical actions involve movement. The technique is designed to ellicit a direct response over the area being massaged, treating identified areas of muscle tension, scarring or adhesion where there is a negative impact, perhaps affecting mobility and creating pain. A good example of this is the gentle stretching of a muscle that has shortened due to injury, assisting re-establishment of length in the muscle fibres, which will help restore muscular balance and ease any stress over the affected joints (see table overleaf).

The body's natural pain suppressants

One of the best-known effects of massage is the release of the body's endorphins; a type of neurotransmitter, a part of our nervous system that controls and sends signals between a

Reflexive action	The body responds to the 'touch' by triggering the sensory nerves, which influence changes to the nervous system, via the brain, that will affect the surrounding tissues and also the condition of other systems within the body. For example, using massage for relaxation eases muscle tension and aids digestion
Mechanical action	The body responds to the pressure and movement applied to individual muscles within muscle groups to physically change the condition and shape of the surrounding tissue through manipulation

nerve cell and other cells. Endorphins also act as pain inhibitors, and can project a feeling of euphoria and well-being. Because of our similar physiologies, it is assumed that the canine has the same response as humans, often demonstrated immediately after treatments via an enhanced joie de vivre! Techniques that can help produce this reaction include effleurage and petrissage.

Massaging to enhance mobility

Lameness in dogs is often caused by poor muscle or joint action, the lameness having a direct impact on the dog's perception of pain. The appropriate application of massage can work extremely effectively where mobility is compromised.

When using massage to treat musculoskeletal problems – mobility problems affecting the muscles and joints – the aim is to promote the dog's natural healing mechanisms. Canines possess highly effective and efficient curative properties within their physiology, processes which are facilitated if the damaged area is presented with the correct healing environment. Massage can help promote this situation by positively influencing the circulatory system, encouraging arterial blood flow to the damaged muscle fibres suffering from poor circulation. This allows enhanced delivery of oxygen and other vital substances, giving the opportunity to repair and re-establish cellular balance and resume normal function.

Massage can really help improve mobility problems in your dog.

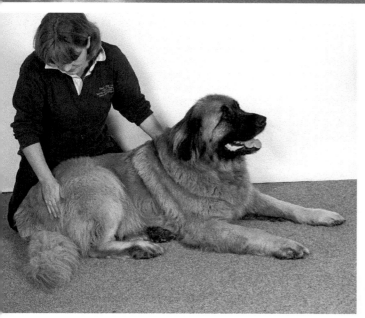

Massage can help promote healing by positively influencing the circulatory system.

Massage also has a direct affect on the configuration of superficial muscle fibre alignment. When the fibres have become congested or shortened, the application of stretching techniques, such as petrissage, can help to restore the muscles' natural functional length.

The working and performance dog

Massage can be extremely beneficial for the active or working dog; a lively companion, a

An x-ray showing a severe case of hip dysplasia (hip joint not fully located in the socket).

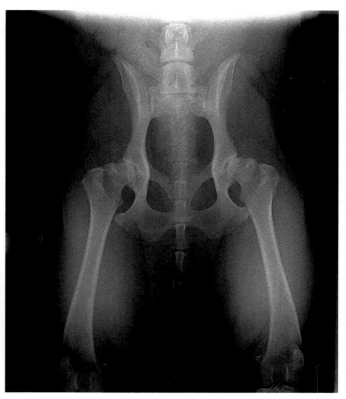

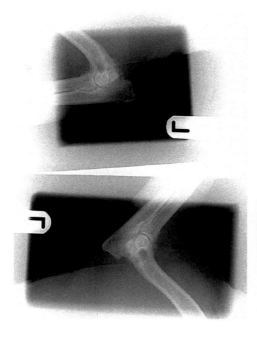

An x-ray showing arthritis of the elbow joint (highlighted by the 'furry' nature of the joints).

Non-specific lameness and general stiffness

Arthritis

Post injury/trauma

(You must seek veterinary advice before embarking on massage in this category)

Hip dysplasia

Post-operative treatment

The support of more complex issues, for example:
Hip/elbow dysplasia
CDRM (chronic degenerative radiculomyelopathy)
Cruciate issues
OCD (osteochondritis dissecans)
Spondylosis

(You must seek veterinary advice before embarking on massage in this category)

CONDITIONS THAT CAN BE IMPROVED THROUGH THE USE OF CORRECT MASSAGE TECHNIQUES

By influencing muscle function through adjusting the length and integrity of the muscle within muscle groups; this will also beneficially influence joint action

By easing muscular tension surrounding the effected joint(s); this will ease pain perception of friction caused by the arthritic joint

By assisting realignment of muscle fibres to help prevent inappropriate healing and shortening of fibres that create balance and compensation issues

By easing tension created through imbalance, and thus helping to reduce compensatory issues. With specific treatments to aid joint support through muscular realignment

Assisting venous and lymphatic return through appropriate techniques that will aid support of the two systems if the dog is recumbent and not mobile, or on cage rest
Assisting venous return which will aid removal of anaesthesia residue
By assisting muscle repair and resumption of activity
By easing muscle tension as a result of a limb being held in place in an operation

These conditions are primary issues that have secondary and subsequent muscular, joint or neural conditions. All can be eased through gentle massage, but for more extensive help seek a professional therapist

A dog and handler running well after being properly warmed up.

competitive dog that participates in events such as agility, flyball, obedience or canicross, or a busy working sheepdog, police or assistance dog. With such animals, besides remedial or therapeutic objectives, massage can also have three completely separate and equally important applications:

1   The warm-up: this is best applied after you have physically warmed your dog (by gentle walking, trotting in a straight line, etc) to enhance

MUSCLE DAMAGE

When an injured muscle is healing, the sliding filaments that make up the muscle fibres (see the chapter entitled How the dog works) will invariably reform in a knotted configuration, which prevents the sliding mechanism from working properly, rendering the damaged fibres ineffectual. This 'knotting' also precludes good blood flow and impedes natural healing. Furthermore, the healed muscle subsequently becomes scarred and is shorter in length, resulting in more stress to the articulating joint and causing imbalance in that area

In this situation, the use of massage can assist by spreading the muscle fibres and enhancing blood flow, so that natural healing and better fibre construction can be resumed. In addition, the same fibres can be gently stretched, aiding restoration of muscle length and helping to re-establish the integrity of the affected joints

Similarly, muscular tension within the body causes pain and imbalance via another convoluted process. The inhibited action of the tense, sore muscles puts stress on joints and other muscles, which, in turn, creates pain, causing further stress

Due to the ensuing damage of these muscle fibres, the mineral balance that creates the correct conditions for nerve stimulation of muscle fibres to occur will also be compromised, leading to problems with muscle tone and function. Contributing to the pain will be the subsequent inflammation caused by additional joint and muscle stress; this will cause further injury or damage to existing or new areas through engorgement, and therefore added tension

Massage can enhance performance of the most-used muscles during all types of activities.

Massage can also enhance delivery of blood to muscles to help maintain activity levels.

In the working canine, such as this sheepdog, massage can assist a pre-work warm-up to help prevent injury.

performance of the most-used muscles. It can also be used when a physical warm-up is not possible; for example, with a working dog.

2 The warm-down (also known as the cool-down): by applying a specific massage after exertion and after a physical warm-down, you can help your dog's body restore homeostasis by assisting removal of any metabolic waste from the muscles. In other words, waste that the body produces as a result of working in an anaerobic state, or when the dog has been panting because of exercise. It is also useful in assisting repair of minor micro-tears within the muscles, which are believed to cause a high degree of post-exercise stiffness. Applying this technique will also provide

CONDITIONS THAT CAN CAUSE CONTRAINDICATIONS IN THE DOG

Condition	Why it is a contraindication
Fever or virus	Massage causes discomfort, and could compromise lymphatic system function by inappropriately enhancing blood flow. Thus, healing could take longer, or even possibly prevented
Immediately before or after the dog has eaten	Massage will promote enhanced blood flow to the muscles and away from the digestive system. Therefore, do not massage one to two hours either side of feeding, in order to rebalance circulation
Skin infection	Massage can spread the infection
Open wounds	Massage can be painful and may prevent healing
In cases of cancer (unless advised by a veterinary surgeon)	Opinion is divided. However, it thought that using massage for palliative care as a way of easing stress far outweighs any negative response
Shock – clinical	Shock is like fainting in human terms; the body simply collapses due to exertion, illness, or some other physiological problem. The collapse occurs because the body has redirected blood to the vital organs and brain to ensure survival. Therefore, redirection of blood to the muscles can have a highly detrimental effect
Straight after extreme exercise or when the animal is not in a state of homeostasis	Before massaging, allow the body to rebalance naturally – especially with regard to oxygen deficiency
Severe oedema/ bruising	Massage can add to the damage and will be extremely painful
If your dog has been involved in a possibly serious accident	It would be dangerous to apply any influence through massage before knowing the extent of any injuries
If you are concerned about your dog's physical condition	If you are uncertain about any aspect of your dog's condition, have this checked by your vet before you attempt massage

an opportunity to feel for any abnormal heat in muscles and joints caused through minor injury. (Note: warm-down massage must be used only when the dog has regained his normal, or resting, heart and breathing rate.)

3   The in-between: will keep the muscles supple and functioning with good range of movement

to enhance performance and help reduce the possibility of injury.

But perhaps equally – if not most importantly – massage can create an even closer bond

A balanced dog is a fit dog.

between handler and dog which will enhance their working relationship and general understanding of each other.

When not to massage

Massage can impact on every system in the body, and there are situations when it is not appropriate, or can prevent quick healing, and even do serious harm. These situations are known as contraindications.

Before embarking on any massage technique designed to enhance your dog's health or mobility, you must always check with your vet that the dog's condition or injury is amenable to massage.

The table below illustrates some of the contraindications that should be adhered to. Please note, however, that this is not a comprehensive list.

CONDITIONS THAT CAN CAUSE CONTRAINDICATIONS IN THE DOG

Condition	Why it is a contraindication
If your dog does not want a massage	For many dogs, massage is an enjoyable experience, but if the prospect appears to make your dog nervous, check out the reason. There are various possibilities; for example: • Your technique – it may be too heavy, too fast, too soft – or tickle! • Your positioning – check this out in the chapter entitled *How to massage your dog* • Your dog experiences discomfort or pain that should be further explored • Your dog is feeling/has been sick and does not want to be handled. If the condition persists, see your vet

4

HOW TO MASSAGE YOUR DOG

When you massage your dog, it's important that you both gain from the experience. The session should be both relaxing and beneficial for your dog, but also equally so for you.

Before starting, ensure you have everything set up properly for the session. Choose a quiet place where you won't be interrupted, and your dog won't be distracted, or choose a time when no one else is around. This is important, because by concentrating fully on what you are doing, you will give your dog so much more, as well as gaining from the experience yourself. Working in the correct environment will also ensure that you can react and respond appropriately to your dog, which will amplify the benefits.

There are two fundamental rules to remember whenever you massage your dog:

1   Do no harm
2   Finish on a good note

Before you start

Ensure your dog is in good health; never massage a dog who is clearly unwell – for example, with

Find somewhere to massage your dog where he feels both comfortable and secure.

a fever or digestive problems. Such illnesses are described as contraindications, and indicate situations when massage should not be carried out. If your dog has an ongoing condition, such as a heart complaint, check with your vet before you consider any form of massage (see contraindications in the chapter entitled *How does massage work?*).

Never massage directly after or directly before feeding; always leave at least two hours each side of feeding time. This is because massage influences the blood to flow towards where your hands are working: in other words, the muscles. The dog's body reacts quickly in this situation, with the result that there is sufficient blood flow away from the digestive tract to possibly compromise proper digestion. This can lead to many complex problems, so is a rule that must always be obeyed. (It is for the same reason that humans should not exercise immediately before, or straight after, eating.)

Whatever technique you intend to use, always begin the massage session with effleurage (see page 47).

The right environment for massage

The room should be an environment in which your dog feels safe, with an ambient temperature of around 20-23° C (68-73° F). Next, consider the best way of working for both you and your dog; If possible, work on the floor with your dog, although you can massage small dogs on your lap. Consider the comfort and safety of both yourself and your dog before you commence. Make sure you are sitting or kneeling in a way that is comfortable, and gives you the room to move and massage effectively: don't forget that you could be in the same position for up to 30 minutes. Also, remember that your dog may wish to stretch out, roll over, or just move about a little; by restricting space you could very easily break the rhythm of the treatment. Your dog should not feel trapped or penned in by the experience.

going to do right at the start: we call this 'intent.' By informing your dog about what you are going to do, especially at the beginning, you will foster a relationship based on honesty.

Your hand position

Begin the massage by placing your hands on your dog in a position and on an area where they are comfortable for you both. Generally, over the shoulders is a good place to begin, or sometimes

At other times, it may be better if your dog is lying on your lap. Consider the comfort and safety of both of you before you begin.

You can massage small dogs on your lap. For some massage, it is useful if the dog is sitting up.

Your approach

During a massage session your approach must be one of patience and calm. It is intended that this be an enjoyable and relaxing experience, and not one that must be squeezed into a swift, five-minute workout, otherwise, you will negate any benefit. You might also consider using a special soft rug or some form of bedding for your dog to lie on. In this way, your dog comes to associate the object with massage, as well as enjoyable, one-to-one time with you. Whatever your approach, be sure to make it clear to your dog what you are

Tell your dog what you are going to do before you begin.

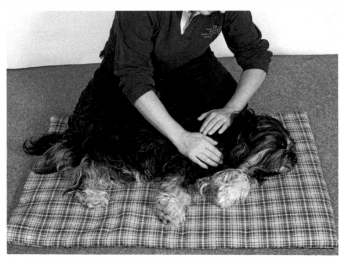

Make sure you are sitting or kneeling in a way that is comfortable, and gives you the room to move and massage effectively.

on the dog's head if he seems happy with the situation. Just hold your hands in place for a few seconds before you commence effleurage; this will provide a physical introduction as well as a clear intention to your dog. Whatever technique you are applying, you must position your hands correctly to facilitate the right support, pressure or movement.

Human 'patients'

Before massaging your dog, first try it out on a willing human 'patient.' Begin by massaging their shoulders or arm; this will enable you to receive verbal feedback on how your pressure feels, both from a fluidity and weight perceptive. When you are working on your human patient, ensure that one hand is resting on their shoulder while the other is working. Next, continue to massage the shoulder with the working hand while lifting the non-working hand off the shoulder. This will demonstrate what it feels like to have the working hand in and out of

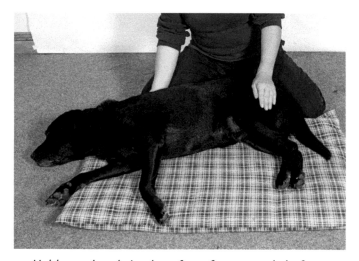

Hold your hands in place for a few seconds before you commence effleurage.

contact, as the hand lifts away from the body after each stroke. This staccato-type massage will be as unsettling and unbalancing for your dog as it is for a human.

Effleurage

In effleurage, it is extremely important to apply pressure evenly over the body of your dog. Your fingers should be closed, and leading your thumb in a very relaxed manner. It is very easy for the fingers to spread, which will result in irregular pressure over the area you are treating. Ensure your wrists are relaxed by shaking them out before you lay your hands on the dog.

The correct technique involves creating a flow over the dog's body in regular and rhythmic

Your fingers and thumb should be together, with your fingers leading your thumb in a very relaxed manner, as shown here.

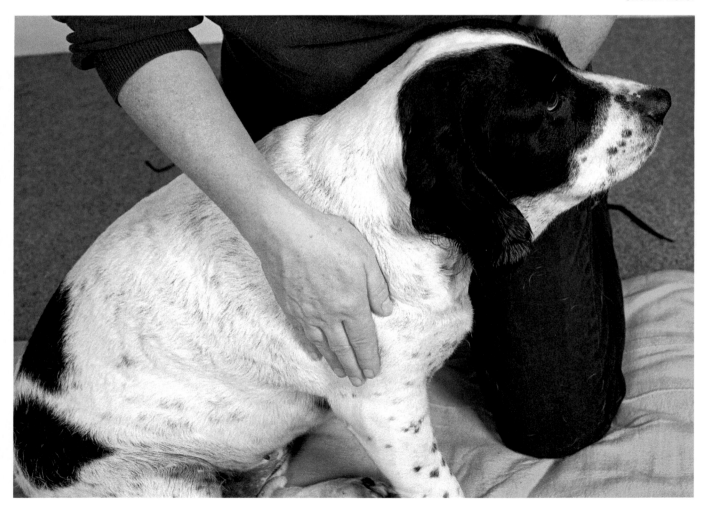

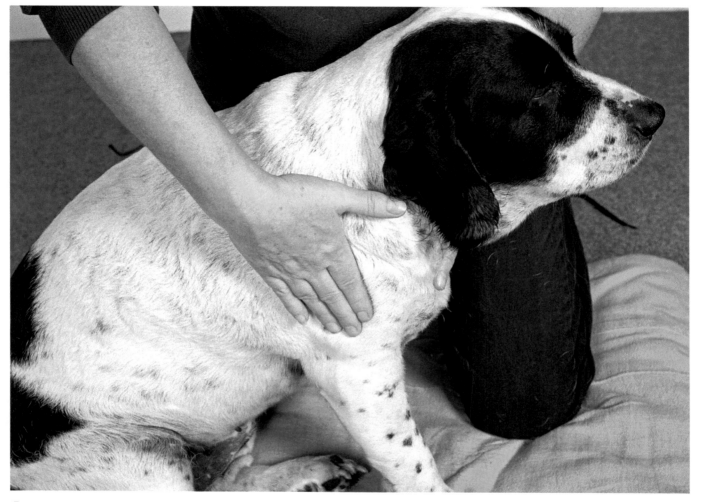

Do not separate your thumb from the fingers, as shown here. Also, ensure your wrists and fingers are not stiff before you start, since this will result in uneven pressure on the dog.

movements, maintaining the same pressure throughout. If you are at all unsure about your technique, it's advisable to have someone watch you work: good pressure, rhythm and flow can be mesmerizing, and your helper can judge whether you are achieving this effect. Better yet, try out your techniques – incorporating pressure and

rhythm – on a human subject before you attempt them on your dog!

Petrissage
Petrissage is a very mechanical movement. Your

hands should gather and move the skin and the superficial muscles in a wringing or kneading action, so should again be relaxed and must not pinch the skin. When applying the 'plucking' technique, your wrists must be really warm and supple to successfully create the rhythm that produces the correct effect.

Passive movement

During passive movement, your hands must be relaxed to ensure you are supporting your dog's legs without gripping or pulling them inappropriately. It is especially important to follow precisely all of the instructions given in the chapter entitled Massage techniques.

Effleurage – how much pressure?

It is difficult to say how much pressure is needed in effleurage, because each dog requires or enjoys different degrees. Much depends on the type of dog and, of course, most importantly, whether or not it has any underlying tenderness in the area you are working on.

It's not true that large or giant dogs prefer more pressure, or, conversely, that smaller dogs prefer less; some Chihuahuas enjoy firm pressure and some German Shepherds can be quite sensitive! It also depends on the individual; if your dog has hip dysplasia, arthritis, or any other form of lameness, its hindquarters and back will be tender. Therefore, a gentler touch may be required there than in, say, the shoulder region, where more pressure is appropriate.

Even if a dog has an underlying problem, light, tickly techniques can be extremely annoying, and won't help engage it in the process.

A useful rule of thumb is to consider effleurage as a 'deliberate stroke;' therefore, apply slightly firmer pressure than you would normally when stroking your dog. You may also be directed by

BASIC MASSAGE ROUTINE	
Start	Introduction; hands placed on dog indicating 'hello'
Effleurage	Warm-up session; essential before any other techniques are applied. This can also be used for an entire routine without other techniques being incorporated
Petrissage	As required; wringing/kneading/skin rolling/pecking techniques
Passive movement	As required; use only when all the surrounding muscles have been warmed and the dog is relaxed
Effleurage	Gentle session to end
Finish	Hands placed on dog for a few seconds and then removed, indicating 'goodbye'
	Stages 1 and 2 must be used at the start of every session Stages 5 and 6 must be used to end each session Other stages can be placed in-between the start and finish

your dog, as he will lean into your hands when asking for more pressure, or away from them when less is desired.

When applying effleurage, always start gently. Then, when you have developed a rhythm, and your dog has engaged with you and what you are doing, you can perhaps try a little more pressure.

Petrissage – how much pressure?

When kneading or wringing it is important not to grip or push too much. This is a technique that requires very little pressure – just the correct application. When plucking, check that you are sliding and releasing the hair and not pulling.

How long?

It's probably best to set aside about 30 minutes for a complete treatment, but ultimately your dog will decide how long he wants the session to last: sometimes it may only be a few minutes, and on other occasions no amount of time is sufficient!

The important point is to respect your dog's wishes and not to inflict techniques on him. If your dog is uncharacteristically fidgety or unwilling, this can indicate an underlying problem – massage can be a good indicator of your dog's general condition. Refrain from chasing your dog around the room if he does not settle immediately. Sit, wait, and encourage him to come back to you. If he still refuses to comply, abandon the session and try again another day.

How to finish

If you are initiating a close, it is always best to finish with effleurage. At the end of a stroke, stop, and then hold your hands still over a comfortable area – for example, where you began the session.

Keep your hands in the same position for a few

seconds, then remove them and sit back to let your dog know that the session has ended.

Changes and findings

In time, you will develop a greater knowledge of your dog and begin to understand more about what is lying beneath his skin. This, in turn, will give you a greater insight into your dog's health. To help you use the information, a good discipline is to record your findings on a simple record sheet, noting any changes or other findings to help you plot your dog's health on a regular basis. It will also benefit your skills of palpation and technique application.

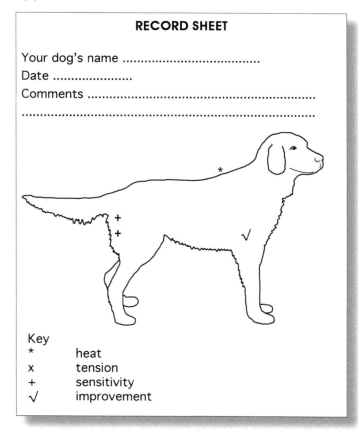

RECORD SHEET

Your dog's name ....
Date ....
Comments ....
....

Key
*   heat
x   tension
+   sensitivity
√   improvement

Using a chart

An easy way of keeping a home record is by filling in a simple chart like the one shown here (see page 128 for template). First, draw or trace the basic shape of your dog as accurately as you can. Indicate those areas where you can detect abnormal heat, tension or sensitivity. You can then refer back to this on your next session, or use this if you decide you should investigate further with your vet and myotherapist.

Massage as a detection tool

Massage is a good way of assessing your dog's health as well as enhancing it. If your dog has a condition that is being treated by a vet or canine therapist, any information you can give them, such as behavioural changes, heat detected within muscles or joints, and whether it occurs only during massage or is ongoing, may prove invaluable during treatment. Therefore, in your massaging sessions, be ready to record any findings so that you can provide accurate information should it be required. Some examples of useful observations are listed below.

LUMP DETECTION

While running your hands over your dog when using different techniques, you will begin to feel areas of tension, or may even find lumps you would not otherwise have discovered. Have these checked out by your vet. Sometimes, the smallest lumps can be the most insidious ones, and it's best to get these examined at the earliest opportunity. Likewise, when ticks are around, their swift detection and correct removal is also beneficial.

COAT QUALITY AND LIE

It is possible to identify different textures within your dog's coat that can indicate an underlying issue. An isolated muscular problem can sometimes affect blood distribution to the skin in the area. Therefore, the coat, being the mirror of your dog's health, may contain areas of dry hair that are inconsistent with the rest of the coat. Also, check to see that your dog's coat is lying evenly; if not, this can indicate stress lines, so, again, make a note and observe any change.

HEAT AND COLD DETECTION

When working on your dog, you may notice areas of heat: either consistent or rising; you may also feel cold areas. All of these can indicate a problem and should be noted. Any area that is constantly hot or cold requires further investigation by your vet, and then possibly by a specialist therapist. If heat is detected when you are applying different

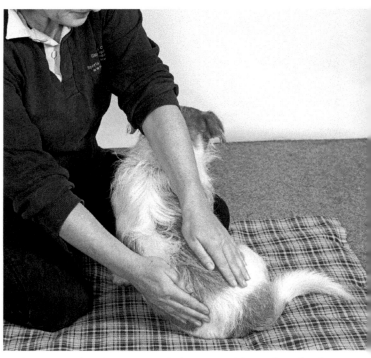

Any area that is constantly hot or cold requires further investigation by your vet, and possibly a specialist therapist.

techniques, make a note, then check to see if it is still discernible when you next use the same technique. You may find that it does not occur again – generally a good result!

SKIN PLIABILITY

The technique of skin rolling is highly indicative of the condition of the skin in relation to the superficial muscles lying below. Skin can adhere to the muscles and fascia below the skin through repetitive strain issues or injury – for example, excessive jumping down and either constant or intermittent lameness. Note where these issues are located, and then gently work over them

to encourage the slow release of the skin from the underlying tissues. This can greatly enhance mobility and also improve skin and coat quality, and is another tool to use for detection of any possible problems.

RANGE OF MOVEMENT

Your dog's level of mobility will be indicated by any resistance during passive movement. The technique will also demonstrate individual limb mobility. This is a good indicator of improvement or decline in joint integrity.

Skin should be loose and mobile.

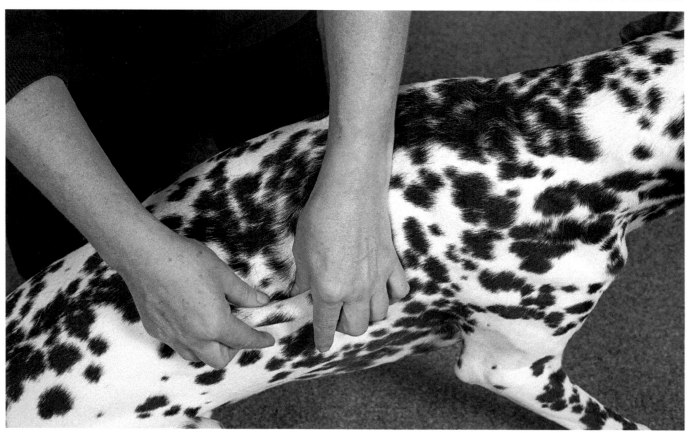

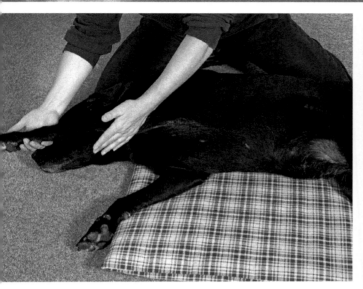

Passive movement, done correctly, will indicate the degree of limb mobility.

After massage, restrict the walk to ten minutes, preferably on the lead, if any issues have been discovered.

After massage

Although you should not feed your dog for a couple of hours after massage, you should encourage him to drink and to go out to evacuate both bowels and bladder, after which, allow him to rest and sleep. This is important, because of the changes you have made to the systems, and the body will need to rebalance as a result of the positive changes that have been made. This can only be properly achieved during rest:

it is extremely difficult for the body to heal and be active at the same time!

Provided you have not detected any heat during massage (as described above), and your dog has enjoyed a relaxing and therapeutic massage, let him rest for a couple of hours before taking him for a short walk (no more than 30 minutes). If you have found heat, tension or any other issues, however, wait for at least two hours before exercise, and then restrict the walk to ten minutes, preferably on the lead.

Do not feed your dog for a couple of hours before or following massage.

Follow up

The frequency with which you massage your dog is dependent on several factors:

- If your dog is happy, compliant and well: once a week

- If issues exist – if your dog has arthritis, for example: no more frequently than every ten days

- If issues have arisen during massage: depending on your dog's reaction to the therapy, if he is happy and enjoying the techniques, then once a week; if he is fidgety but still absorbing the treatment, no more frequently than every ten days

- How much time you have: being calm and relaxed and not working to a tight schedule will give most benefit

- The age of your dog: generally, if your dog is eleven years or older, it is best to leave at least ten to fourteen days between treatments, since it takes an older body a little longer to make any changes

My dog does not settle during massage	**TROUBLE-SHOOTING GUIDE**
	• Ensure your dog is not suffering from any illness (see contraindications in the chapter entitled *How does massage work?*)
	• Check your pressure and grip – maybe too much or too little?
	• Check your rhythm – it may be better to apply effleurage with one hand following the other (see instructions on application of effleurage)
	• Check that your non-working hand is in constant contact with the dog
	• Try plucking – this can be surprisingly relaxing – or even skin rolling
	• Check that you are not making your dog feel insecure or dominated by leaning over it too much
	• Your dog may have an underlying issue of which you are not aware, and is worried about you causing pain by touching or putting pressure on the area
	• Your dog needs to go out! Massage is highly stimulatory, which can have a direct effect on the bladder and/or bowels
	• Your dog's routine has been interrupted and it's suppertime or walk-time!

My dog settles for a while, but then fidgets

My dog quickly glances at a particular area I am working on

I feel heat where I have been massaging

My dog will not let me apply passive movement

What if I find a very sensitive area on my dog?

TROUBLE-SHOOTING GUIDE

- See notes 1-6 opposite

- Establish whether your dog moves after, or during, a particular technique. Check also to see whether he moves when you work on a particular area of his body. If he does, check your pressure, or be more aware of which area appears tender

This usually indicates sensitivity in that area. Work more gently here to gain your dog's confidence. By doing this, you may be able to help by applying techniques that ease any congestion. If the sensitivity persists, seek veterinary assistance

- If the heat develops during the massage, this generally indicates a good reaction caused by corrective blood flow, possibly easing congested muscles. The heat should dissipate within minutes and not cause your dog any distress: in fact, your dog will probably find the experience a pleasant one, and it could even be that massage will ease an underlying mobility problem. If, however, the heat persists, this is an inflammatory response, indicating an underlying problem that should be investigated

- Heat is commonly detected during skin rolling, especially if over an area where the skin feels tight in comparison with similar spots. The heat can indicate a highly positive effect, but be careful not to 'over roll' the area, and be aware of adjacent spots that may be slightly stuck and therefore very painful for skin rolling

This usually indicates a mobility issue, so great care must be taken. To gain confidence, try applying the technique using almost no movement: hold in place, then release. Remember that you are capable of applying a force that can override his own. Therefore, he must feel confident that you will not do this

Do not force any technique on the area. Try using passive touch to gain your dog's confidence; this will ease tension slightly and could lower pain perception, thus allowing you to try a little effleurage. If that doesn't appear painful, revert to passive touch. If the area remains sensitive, have it checked out by your vet

TROUBLE-SHOOTING GUIDE

During a massage session, my dog goes into a very deep sleep. Is this normal?

It is so rewarding, yet also slightly unnerving, if this happens. However, it is generally a very good sign that your dog is really benefiting from both the treatment and the security he feels. Dogs seem especially sensitive to the endorphins (pain-relieving chemicals) they produce, and these can change a really active dog into one who is completely switched off!

My dog goes and hides under the table when I start to massage

Resist the temptation to try and chase your dog from his hiding place, and instead encourage him out by opening your hands and calling. Your dog may be trying to demonstrate 'displacement behaviour' – in other words, trying to distract you from what you are trying to do for some reason. It is possible that he is nervous and unsure about a new routine, feeling sore, or just very insecure. Do not try any form of treatment until your dog is ready to accept it

Why does my dog roll onto my hand when I massage him?

This can happen when a dog wants you to apply pressure to a particular area. It can also mean that he wants you to leave the area alone. You must judge which of these is the likely reason. Try applying effleurage to the area when he is sitting upright; if he leans towards your hand, he is probably asking for more pressure; if he moves away, then the area is too sensitive

My dog yawns during treatment

This has a variety of different meanings, depending on what the rest of your dog's behaviour is indicating. If he is standing and yawning, this could indicate that he is rather apprehensive. If he is lying down and relaxed and yawning, this could indicate that he is happy and accepting the treatment

Why does my dog always present his neck/bottom to me?

This could be your dog's way of trying to show you which part of his body he wants you to work on

5

MASSAGE TECHNIQUES

There are many different massage techniques that can be used on dogs. Each technique has a different effect, designed to produce a specific outcome for a particular problem. What is common to them all, however, is that they must always be used sympathetically and, as previously mentioned, on a dog who is happy to be treated in this way. Most techniques have a set of 'rules' or guidelines that must be adhered to so that you never harm the dog: refer to the chapter entitled *How to massage your dog* to check you have created a conducive atmosphere, your dog is feeling well and prepared, and that you are also relaxed.

Another extremely important aspect of any massage is that, throughout the whole session, you must always keep one hand in constant contact with your dog's body. This creates a 'balance' and gives continuity to the technique.

Effleurage

This is the 'stroking' technique that is the foundation of most massage treatments or sessions. Application of this technique is generally the first point of contact with a patient, as well as your initial introduction, and because of this it is vital that the application is correct and appropriate. To gain your dog's acceptance, it's important to apply the correct pressure, together with a good rhythm. If you can establish a good rhythmic stroke pattern, especially initially, your dog will be more inclined to engage with the technique.

HOW IS EFFLEURAGE APPLIED?

Your hands should form a soft and uniform shape, so that the pressure exerted through them is even from fingertip to palm. This is important, because pressure must not be applied through the fingertips only. If this happens it will cause discomfort to your dog – preventing any form of relaxation or engagement with the therapy – and can even cause injury. (This tends to occur more with smaller

The uses of effleurage

- It acts as an introduction to massage
- It helps to relax your dog
- It creates the first bond through 'touch' that will help the dog to understand that massage makes it feel better
- It warms the muscles to facilitate petrissage or passive movement
- It can 'link' different techniques or sequences – for example, when applying passive movement, you can effleurage the leg between each section
- It warms the muscles that will help prepare your dog before an event or exercise
- It can help you assess areas of excessive warmth or cold
- It can help you assess areas of muscle tension or sensitivity
- It indicates a finish to the session by gently settling nerve endings and muscle fibres

dogs, since it is more difficult, due to the smaller scale of the dog and the available areas for application, to apply even pressure using only part of your hand or fingers.)

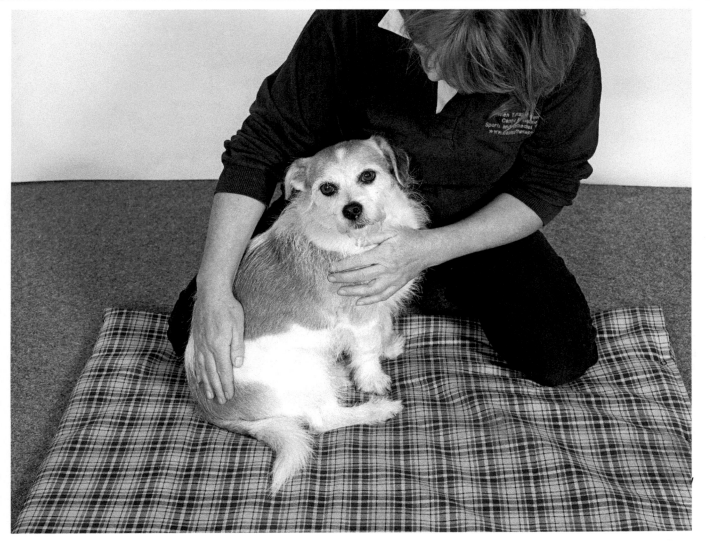

During effleurage, your hands should form a soft and uniform shape.

The first step is to have relaxed hands, so gently shake out your wrists before you begin. One of the main reasons for dogs accepting a stroke, but not necessarily a massage, is because a stroke is delivered with a lightness and flow of the hand, whereas when a novice begins to massage, the hands and wrists invariably tense up. This physical tension causes the massage technique to have an uneven flow, creating uncomfortable pressure points for the dog. Under these conditions, the

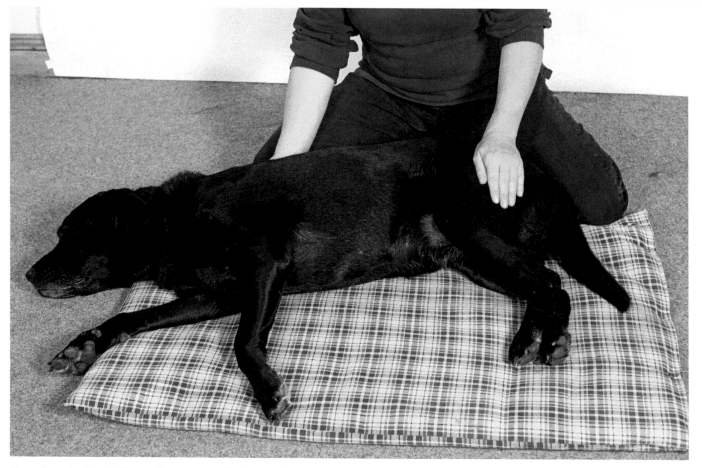

Your hands should form the same soft shape, irrespective of the size of the dog.

dog will find it very difficult to settle and enjoy the massage.

HAND POSITION
Keep your hands in a soft, flat 'mitten' position rather than a 'glove' position – in other words, keep your fingers together and your thumb following your fingers. Be especially vigilant if your dog has a long coat, since it is easy to allow your fingers to

spread and therefore not exert even pressure.
Effleurage can be applied with one hand, or with two hands in a 'one hand following the over' rolling technique. Sometimes, using two hands is an easier way of establishing a good rhythm.

STROKE LENGTHS
The strokes generally feel more relaxing when they run the full length of the muscles, or when they flow

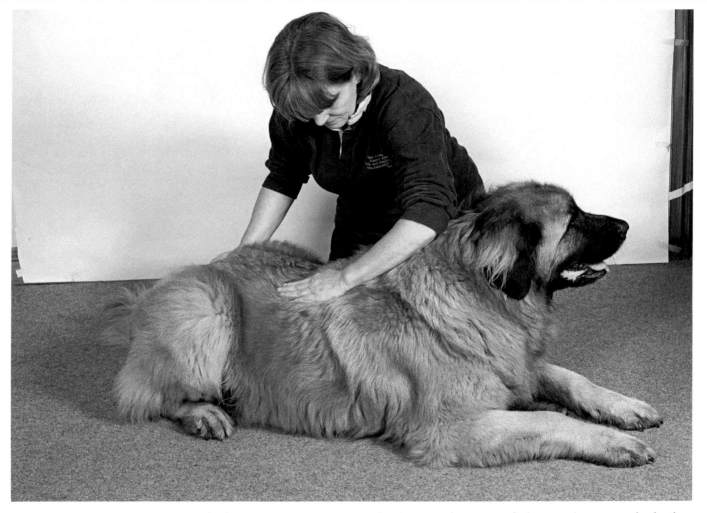

The best way to engage your dog is to apply your techniques using a steady rhythm.

all the way down the neck and shoulders, or all the way down the back or legs. Again, this provides a continuity of approach, and your dog will be able to anticipate where the stroke is going, and therefore begin to work with you, and maybe even lean into your hand!

RHYTHM
The best way to engage your dog is to apply your techniques to a rhythm; particularly with effleurage, because it is usually the first technique used. Try to use a regular rhythm as well as one that is intended to achieve a specific outcome. For example, if

61

you are seeking to relax your dog, apply long, slow rhythmic strokes; if you are using the technique as a warm-up, use long and rhythmic – but faster – strokes (although not harder ones).

PRESSURE

Start off initially with positive, but not hard pressure. Think of it as a 'deliberate' stroke; thus, it can be a little firmer than a stroke. If your dog seems happy with the pressure, then continue. It can then be applied a little more firmly, subject to your dog's approval, when you have been applying the technique for about five minutes all over. You may find that you cannot get your dog to settle, however; if that is the case, try softer pressure or very slightly firmer pressure, and also try changing the rhythm.

Your hand pressure should start off as a 'deliberate' stroke.

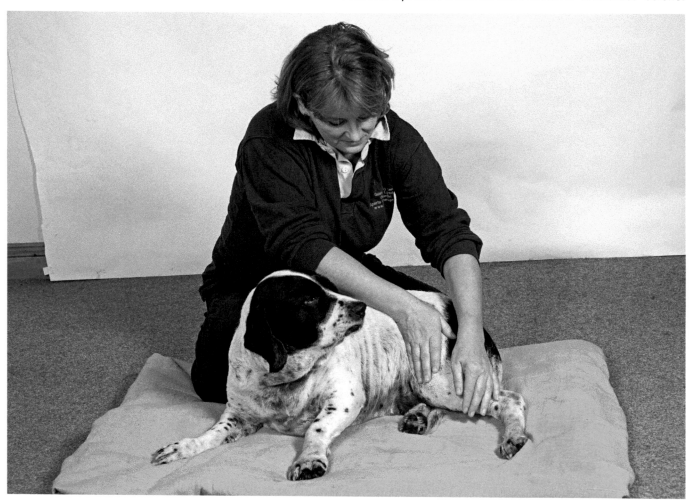

- Technique applications include: one-handed, and hand-over-hand
- Can be applied superficially or deeply when the muscles are warmer
- Applied longitudinally (in the direction of the superficial muscle fibres): this usually corresponds with the way a shorthaired dog's coat lies
- Can be applied to the cheeks, ears, top and sides of the head, trunk, back and hip region, legs and back
- Hand, fingers or fingertips should slide readily and evenly over the dog's coat
- Maintain even pressure for the entire stroke and from one stroke to the next; pressure can slightly increase as muscles warm and if recipient is happy
- Remember when massaging to keep breathing

WHERE CAN EFFLEURAGE BE USED SAFELY?
You can use effleurage on most of your dog's body. The best place to start is on the neck, beginning with your fingers pointing downwards. The technique is best applied with him lying down.

- Apply the stroke all the way down the side of your dog's neck to his shoulder, then onto the top of the leg (1)

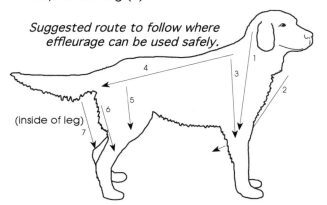

Suggested route to follow where effleurage can be used safely.

- Then travel down the neck again, between the front legs, and on to the chest (2)
- Start from the top of the shoulder and massage down the shoulder to the top of the leg (3)
- Then massage all the way down your dog's back on each side of the vertebrae. Do not massage over the top of the vertebrae. When you massage the back, ensure you massage close to the vertebrae, as this is where the main supporting muscles are situated (4)
- Then start the next stroke from the very top of your dog's back leg (at the pelvis), and massage the whole of the front of the leg (5)
- Start the next stroke almost under the dog's tail, and include the whole of the muscular part of his back leg (6)
- Then, on the other side, go back to the same start point and massage under the dog's leg to end up at the same finish point (7)

Whether you use this or another route, ensure you travel in the directions specified, and observe the start and finish points. You needn't complete the whole routine; your dog may just want his neck or shoulders massaged. If so, don't apply other techniques to any area that requires warming first until you have applied effleurage.

Passive touch

This is a very gentle and non-manipulative technique that can be used over areas of extreme tension or sensitivity. It can also be used on a dog who is not accustomed to being touched.

WHAT DOES IT DO?
It has a calming and warming effect that can initiate the development of trust in a nervous dog.

HOW IS IT APPLIED?
It is applied with no pressure or movement; just a gentle touch using the palms of your hands and fingers positioned in the same way as for effleurage, and placed lightly on the body. Some dogs who are extremely resistant to touch find it less intrusive if you use the back of your hand.

WHERE CAN IT BE SAFELY USED?
It can be used anywhere on the body. Take your time, and leave your hands or fingertips in place for between 30 and 90 seconds. Relax, be calm, and remember to breathe!

Passive touch is a very gentle technique that can be used over areas of extreme tension or sensitivity.

Petrissage

Petrissage is the collective name for a group of manipulative techniques that are intended to gently stretch and move the belly muscle (the main bulk of the muscle) to encourage enhanced circulation. There are a variety of definitions for this technique as well as several categories. However, their names are less important than what they do, and when you should use them.

Gentle kneading

WHAT DOES IT DO?
The technique is to gently 'pick up' the underlying muscle fibres and then release them again in a 'lift and release' routine (see photos opposite). This has the effect of forming a compression followed by a release that can influence both the superficial and deep tissues of the dog's body. By constricting blood flow to the local area and then lightly releasing it, it acts as a system for cleansing the area with fresh arterial blood. This technique begins with very light pressure, which can then be increased as the tissues warm, and your dog becomes accustomed to the procedure.

The technique can help break down adhesions, scar tissue and engorged muscles. If applied correctly, it can have a relaxing, calming effect.

HOW IS IT APPLIED?
By kneading, as you would knead bread dough. It is a deep, rhythmic technique that must only be used when the area has been warmed by effleurage. The technique takes a little practice, so, again, it's advisable to practice on the arm of a human patient first so that you can develop a smooth and even-pressured application, by lifting the tissue without pinching the skin.

Begin the technique by using a hand to cup an area – over the top of the neck is often a good place to start. Then, forming a wavelike action, gently run your hand up both sides of the neck with your fingers. Grip the neck in a moderately

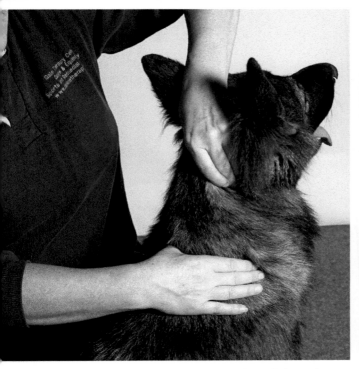

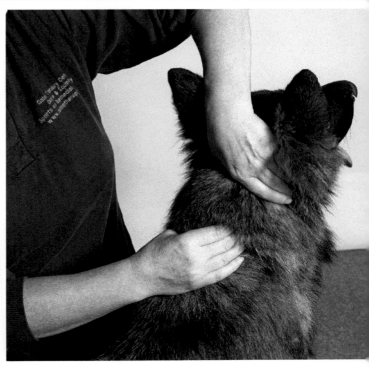

Begin with one hand on top of the neck and the other just above the shoulders. Run your top hand gently up the side of the neck, gripping it as you do so. Run your lower hand, gently gripping, down the neck, working in the opposite direction to your other hand in a wavelike motion ...

... carry on running your top hand down the neck and your bottom hand up the neck in a continuous action.

firm fashion, lift the muscle fibres, and then release. When your hand reaches the top of the neck, start again. Both hands perform the same action, but when one hand is just starting the stroke, the other should be just finishing, creating a rocking technique. It can be used with a few fingers only on smaller dogs, or just one hand.

WHERE CAN IT BE SAFELY USED?
This technique must always be applied after effleurage, so that the dog is prepared, and also so

that the underlying tissues are warmed before they are manipulated. The best places to perform the technique are over the neck and very gently over the shoulders; it's not really suitable for the legs or abdomen.

Gentle wringing
The name of this petrissage technique pretty much describes the action of it. It must be applied carefully or it's possible to pinch the skin and for it to be very uncomfortable (see photos page 66).

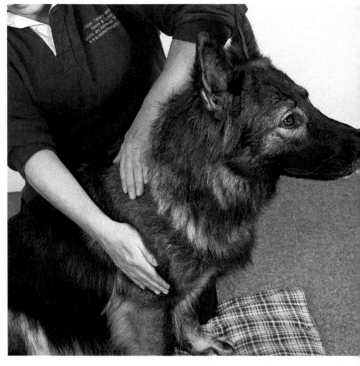

Lay your hands gently over the area to be massaged and, working in opposite directions, slide your hands over the dog (in other words, with the top hand sliding down the dog and the bottom hand sliding upward) ...

... continue this rhythm by sliding your hands up and down in opposite directions.

WHAT DOES IT DO?

This technique is good for separating muscle fibres, and also for mobilizing the skin and underlying tissue. By moving the skin and influencing the muscle below, the technique can also aid venous return, and therefore help break down slight scarring and adhesions. If applied correctly, it can have a relaxing, calming effect.

HOW IS IT APPLIED?

It is applied by sliding your flat hands in opposite directions over the surface of the dog. It can also be used on the legs, but here your hands should

be cupped so that they 'cup and slide' in opposite directions – just as they do when wringing out a cloth (albeit without the pressure). Practise this first on someone's arm, because, if the technique is not applied correctly, it is very easy to pinch the skin.

WHERE CAN IT BE SAFELY USED?

Over the neck and shoulders, and carefully over

Gather a small roll of skin between your fingers and thumb ...

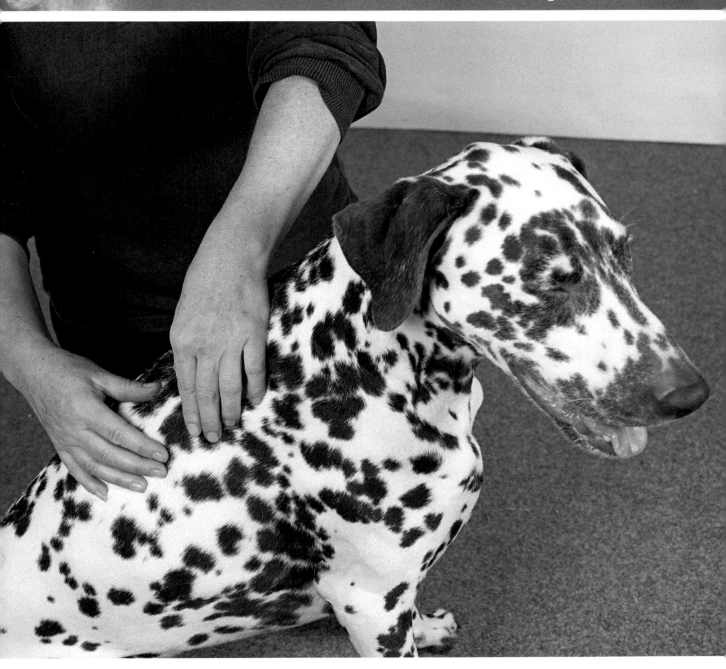

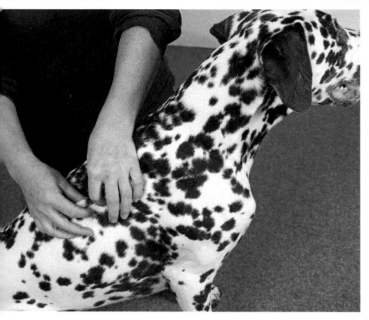

... very gently roll the skin by 'walking' your fingers along it ...

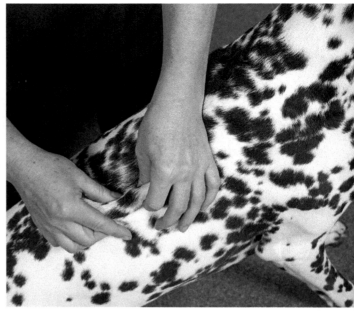

... your thumb should be passive; used only to maintain the rolling 'wave.'

the torso; also around the legs using the cup and slide technique.

Skin rolling
WHAT DOES IT DO?
This helps to remove adhesions caused through chronic injury, which can create stickiness between the skin and underlying fascia that will prevent good mobility of the skin. Adhesions can build up very easily, especially in cases of repetitive strain, and releasing these can improve mobility and greatly reduce discomfort. Skin rolling also aids blood flow to the skin and benefits skin condition, thereby improving the coat. For a dog with no problems within the skin or underlying tissues, this can be extremely relaxing, for recipient and provider alike.

HOW IS IT APPLIED?
Again, it is strongly advised that you practise this technique on a human subject before your dog. Using both hands, gather a small roll of skin between your fingers and thumb (the number of fingers you use depends on the size of dog). Very gently roll the skin, 'walking' your fingers in any direction over the dog's back or shoulders: your thumb is passive, and is just used to maintain the rolling wave. It should be a smooth action, reminiscent of a wave rolling down or across the dog's body. It can be applied in any direction – the more changes of direction, the better.

Great care must be taken if the skin is tight or does not feel loose to the touch; skin rolling over skin like this will feel extremely tender and uncomfortable.

A dog's skin should flow easily with this

technique, although this will vary according to the breed. If the skin does not flow, it is a good indication of an underlying problem, so continue very carefully and sympathetically. It's amazing how quickly this technique can positively benefit skin elasticity.

WHERE CAN IT BE SAFELY USED?

The technique can be used over the neck, shoulders and back. Take care when using it over the pelvis and over the ribs. It's not really suitable for legs.

Carefully gather a few hairs between your thumb and index finger ...

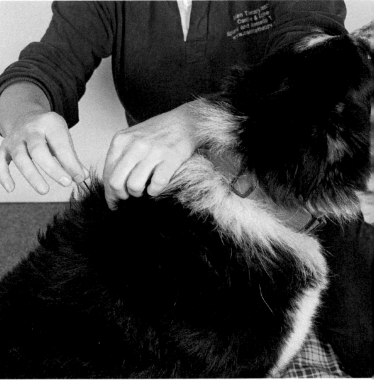

... and gently slide your fingers up the hair shaft.

Plucking

This is an extremely good petrissage technique for relaxing a dog and getting it accustomed to being touched. It is extremely useful if you have a nervous dog. Unfortunately, it will be very difficult to apply successfully to dogs with very close coats.

WHAT DOES IT DO?

This is a reflexive technique (see the chapter entitled *How does massage work?*) that can relax a dog extremely quickly. Mildly tugging the hair produces a very gentle manipulative effect: the canine equivalent of an Indian head massage!

HOW IS IT APPLIED?

Carefully gather a few hairs between your thumb and index finger, and then, very gently, slide your fingers up the shaft of the hair. Repeat in a plucking action so that the tempo is quite fast. (two plucks per second). Continue in a good, even rhythm, taking care not to grip the hair but just slide your fingers over it. Also take care not to pinch the skin

Once your dog is comfortable and stable, preferably lying down, take up your position at his back. Gently take the front leg and, from the elbow, GENTLY extend the leg by easing the elbow forward – do not pull the leg. Support the carpal/wrist joint with your other hand. STOP the moment you feel any resistance – this may be after very little movement. Hold the leg gently in position for ten seconds, then carefully allow it to return to the 'neutral' position.

or pull the hair; the faster the technique is applied, the greater will be the likelihood becoming heavy-handed.

WHERE CAN IT BE SAFELY USED?

Plucking can be applied almost anywhere; the most popular areas are over the neck, back and chest.

Passive movement

This technique moves the legs without 'active' participation from the dog. The technique has extremely strict rules that must be followed even more stringently than the others.

With the leg in a neutral position, apply effleurage again, allowing a little time for the leg muscles to settle.

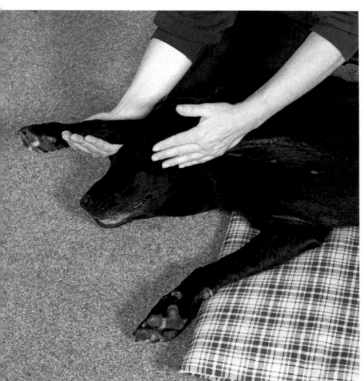

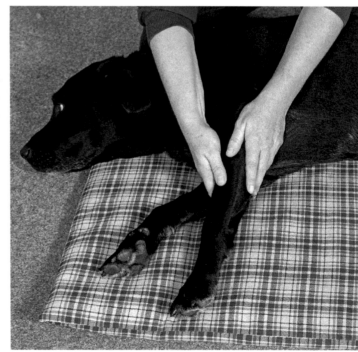

RULES FOR USING PASSIVE MOVEMENT

- Muscles must first be sufficiently warmed by effleurage

- Passive movement is used to take joints through their natural anatomical movement; it is not a stretch

- Passive movement should be conducted slowly, without any sudden or unpredictable movements, so that all of the tissues involved have time to adjust. If a joint is moved quickly, damage to the muscle and joint can result

- The movement should be held in position (flexed or extended) for at least ten to fifteen seconds

- After every movement take the leg back to a neutral point (natural position) and effleurage before applying the next movement

- One passive move is sufficient for each joint

- All moving limbs must be supported above and below the joints

- The dog must be comfortable and stable – either standing or lying

- Passive movement must not be used by an open wound, freshly injured area, any swelling, or on a locked joint

- After passive movement, finish with effleurage/light exercise to readjust joints and tissues

NB: Passive movement can be used on arthritic joints, but only within the range of movement of the particular joint

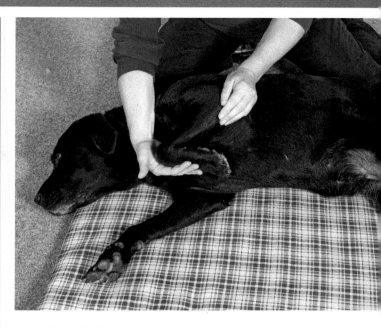

Gently hold the elbow and ease upwards with one hand, while the other hand supports the carpal/wrist joint in a flexed position. DO NOT APPLY ANY FORCE THROUGH THE WRIST/CARPAL JOINT. STOP the moment there is any resistance. Hold in position for ten seconds. Release and allow to return to the neutral position, and again massage the leg muscles.

WHAT DOES IT DO?

Passive movement is important within a massage routine for both the active and less active dog, because it can have a positive effect in maintaining or improving movement range and therefore mobility, by influencing muscle memory.

Muscle memory comes into play when a joint is injured: the muscles surrounding it involuntarily tense or go into spasm as a way of protecting the area from further damage. This is an excellent mechanism when the joint is in an inflammatory situation and rest is essential.

Remedially, however, a joint requires movement to aid natural healing and therefore enhanced

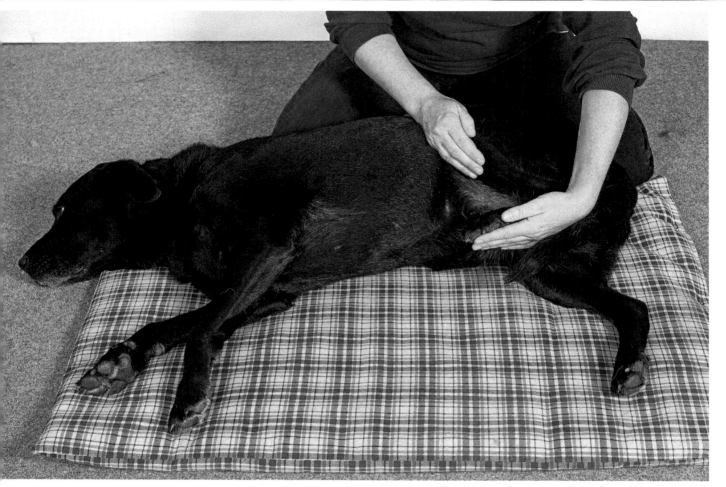

Gently take the leg and 'cup' the stifle/knee with one hand, while placing your other hand underneath the leg over the paw and the hock. With your 'cupped' hand, gently ease the stifle upward, holding the leg against the dog's body. With your other hand, support the hock and maintain the flexed joint. *DO NOT PUSH ON THE HOCK OR LOWER LEG. Take the leg to the position where the dog is comfortable and STOP the moment there is any resistance. Hold for ten seconds. Release the leg by allowing it to go back to the neutral position. Massage and allow for the muscles to settle.*

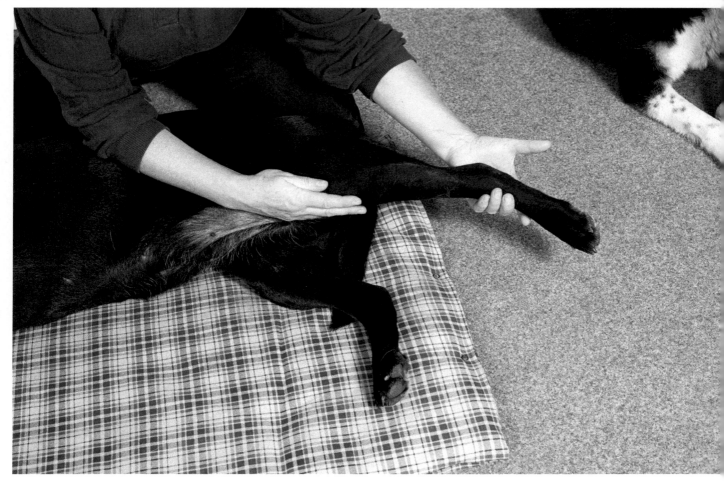

Again, by placing one hand on your dog's stifle and the other underneath the hock, gently ease the stifle toward the tail, taking the leg directly behind your dog, as it would naturally move, and not lifting away and toward the centre line. Ease through the stifle and DO NOT PULL THE LEG; your hand must only support the hock and not apply any pulling action. Hold for 10 seconds in a position that is comfortable for your dog. (The dog in this photo has a fairly good range of movement; yours may only be able to move an inch/few centimetres or so backward.) Release the leg back to the neutral position. Massage to settle the fibres.

mobility. This technique stimulates the neural pathways of the joint by informing the area that there is less need to protect the joints through involuntary muscle tension; by helping this situation and reprogramming how the body views the area or areas, this technique can greatly increase range of movement. This does not just help the dog move more easily and quickly, it also has a positive effect on joint health by promoting an increase in the synovial fluid within the joint. This will help ease the effects of arthritis, reducing pain and improving mobility.

The technique can be incorporated into warm-up and a warm-down routines, as well as in remedial cases.

HOW IS IT APPLIED?

The method involves taking the leg and its joints through their own natural range, then stopping and

THE EFFECTS OF DIFFERENT MASSAGE TECHNIQUES

Technique	Mechanical	Reflexive
Effleurage	Gently replicates muscle movement through slight pressure and slight stretching	Soothes, relaxes, and assists endorphin release. Encourages 'rest and digest.' Influences venous return and therefore arterial blood delivery
Passive touch	None	Warmth of the hand encourages dilation of blood vessels and therefore enhanced blood flow. Reassures, encourages and relaxes
Petrissage GENTLE KNEADING	Lifts and gently stretches the superficial muscle fibres	Encourages vasodilation and promotes muscle healing
GENTLE WRINGING	Moves and gently stretches the superficial muscle fibres	Eases and aids muscle fibre integrity
SKIN ROLLING	Gently stretches the skin and eases any adhesions from the tissue below	Eases tension and relaxes and encourages the parasympathetic nervous system

THE EFFECTS OF DIFFERENT MASSAGE TECHNIQUES

Technique	Mechanical	Reflexive
Petrissage PLUCKING	Gently manipulates the hair follicles	Encourages the parasympathetic nervous system
Passive movement	Eases joints through their natural range and can help venous return through replicating movement	Positively influences joint and muscle memory, encourages the neurological elements within the muscle and tendons to allow the muscles to relax, thus aiding range of movement

holding the movement when the dog shows even the slightest resistance. If during the procedure your dog starts to 'kick' or resist the 'holding' movement, it means that you are taking the leg too far. Ease the leg back towards the neutral position until the dog is comfortable and no longer resists.

Important: this is not a stretch; the stretching of a dog's legs by an untrained individual can do great harm! Please read and apply the instructions carefully.

6

MASSAGE, EXERCISE, and MUSCULAR HEALTH

This chapter explains how massage and hands-on exercises can benefit your dog, from puppy to veteran: whatever the age of your dog, there are many ways you can assist his mobility, help him achieve optimum condition, be warmed up ready for competion work, or ensure he enjoys good muscular health.

The puppy

When it comes to exercising a puppy, there's no 'one size fits all' set of instructions. Even within a given breed, each puppy is built slightly differently, and has his own individual characteristics. And, of course, physical maturity is reached at different stages according to the breed of dog concerned.

How the puppy's body develops

Various studies have shown that the growth plates within the long bones of a puppy (for example, a limb bone such as the femur), do not fully form and harden until the dog is about twelve months old, and in giant breeds it may not occur until the dog is eighteen months old. This means that your puppy's bones retain soft cartilage sections within them until he has fully matured.

Bones require physical stresses to create strength within the shaft, but these need to be appropriate and applied in the correct way. Nor is it just formation of the growth plates that must be considered, as other components of the skeleton must develop correctly if the puppy's body is to mature healthily.

Within the synovial joints (see the chapter entitled *How the dog works*), the synovial membrane nourishes the cartilage that protects the bones and provides concussion. The synovial membrane, a thin layer surrounding the synovial capsule, is a sensitive and delicate structure that can be damaged by over-stressing (excessive pounding or twisting), which will adversely affect the cartilage. The result can be long-term problems with the joint, and thus mobility problems for the maturing puppy.

Surrounding and supporting the synovial joints is more soft tissue: the ligaments and tendons attaching to muscles. The ligaments and tendons maintain a joint's integrity and protect it from stresses; if they are damaged or malfunctioning, this will impact on the joint as a whole. Any injury or overexertion, however small, can affect muscular balance as well as the integrity of working joints. Furthermore, a muscular dysfunction can affect more than one joint, with subsequent compensatory issues. For example, an injury to the front part of the dog can end up affecting back end movement.

Your puppy needs to develop his spatial awareness, and learn how to use both fore- and hindlimbs appropriately to create good movement. A growing puppy also needs to be aware of his own increasing size so that he can avoid accidents – such as when trying to squeeze into spaces that are too small. Appropriate massage techniques and exercise can help to ensure development of good stability, correct muscle patterning (the sequence in which the muscles work), and spatial awareness.

Native puppy versus domestic puppy

A puppy growing up in a domestic setting develops in ways that differ from how he would develop in a natural environment as part of a pack.

By examining the amount and type of exercise undertaken by 'native' puppies as they grow to maturity, we can, perhaps, judge what is appropriate for our own pet dogs.

For example, in a pack situation, a puppy would not go hunting until he was mature (at about nine or ten months old), and not before learning the necessary skills from his parents. Listed here are some of the likely differences between the lifestyles of a natural puppy and domestic puppy, and the possible implications.

Puppies growing up in a domestic setting develop in ways that differ from how they would develop in a natural environment as part of a pack.

These puppies are developing natural movement patterns with their siblings and not overexerting themselves.

These puppies playing in an almost natural environment are still with their mother.

It is so difficult to know if your exercise programme is doing insidious harm. Puppies generally do not instantly become lame or stiff (unless it is acute), so it's difficult to know what is going on under the skin. Therefore, where possible, perhaps the aim should be to replicate for your own puppy the kinds of activities and exercise that he would experience in his natural environment.

Using your hands as a detection tool

A good indication of your puppy's health can be gauged by laying your hands on his body and feeling for heat, or for resistance to touch through tension. This will help you learn more about what is happening below the skin (see the table on page 83). Laying your hands on the puppy's body will firstly accustom him to being touched, as well as giving you the opportunity to appreciate any heat fluctuations, and identify areas of temperature differential: for example, is one side warmer than the identical spot on the other side? This is an exercise that will prove invaluable, not just during puppyhood, but throughout your dog's life. It's very quick and easy to do, and the more you do it the better you will become at recognizing areas of heat.

continued page 82

NATIVE PUPPY	DOMESTIC PUPPY	POSSIBLE IMPLICATIONS
Playing with siblings using lateral and medial movements – in other words, with the forelimbs and hindlimbs moving away from the body and towards the body	Generally walking in a straight line on a lead	By moving primarily in one plane of movement (forward or backward), the muscles that support the forelimbs and hind limbs of the domestic puppy are not 'fired:' in other words, these muscles, which are vital for stability, are not trained to work
Puppies in the litter exercise by varying their pace: running, walking, pouncing and charging	The domestic puppy goes for a walk on the lead, with the pace and gait dictated by his owner	By walking in a set pattern and generally on one side and at one pace, the domestic puppy is not given an opportunity to develop his own pace, change pace, or maybe even travel in other directions (for example, the other way around a field). This lack of variation restricts natural development
Puppies climb over each other	At home puppies are usually walked on the flat	Climbing over soft siblings and their mother helps the puppy's neural pathway development, and encourages spatial awareness, balance and stability
Puppies begin to learn to hunt. They start to use their nose within their habitat, learning lifeskills, realizing their psychological needs and aiding their physiological development	For lively, excitable puppies, exercise is used to exhaust or tire them. Psychological exercises (scenting exercises) are not always exploited to facilitate this. Is the expression "thank goodness my puppy/dog is shattered" a good one?	Thirty minutes of scenting exercises is thought to be equivalent to five hours of walking, as far as stimulation is concerned. It is fun and mentally exhausting without being high impact; it also naturally stretches the neck and back, which helps naturally develop hindquarter engagement

NATIVE PUPPY

Puppies may play in ten to fifteen minute bursts and then rest, building up to maybe 30 minutes as they mature

DOMESTIC PUPPY

We exercise within timescales that suit us, not ones that suit the puppy. For example, we walk the puppy when we are ready, not when the puppy is naturally awake

POSSIBLE IMPLICATIONS

Excessive stresses can damage soft tissue, and this will affect joint and bone development of the domestic puppy

A domestic puppy walking on the lead goes at a pace and gait dictated by his owner.

Place a hand on one side of your dog and place your other hand in the same position on the other side. Either the palm or the back of your hand can be used; if you have a slightly nervous dog use the back. Leave your hands in place for at least 30-40 seconds, and then move them slowly over your dog's body in a systematic manner.

Here's a suggested sequence:

- Begin with your hands on your dog's neck, behind the ears
- Move your hands down over the front of the shoulders
- Put your hands on hi shoulders
- Move your hands down over the elbows
- Move your hands down the legs

- Move your hands back to just behind the shoulders
- Move your hands systematically down the back
- Hold your hands over the hips and pelvis
- Put your hands over the outer surface of the hind legs
- Move your hands down to the 'knees' (stifles), and on down the legs
- Put your hands inside, and behind, the legs

The benefits of puppy massage

There are several distinct benefits that your puppy can derive from good massage.

Feeling for muscular/joint 'heat' is an extremely useful technique for any handler to develop.

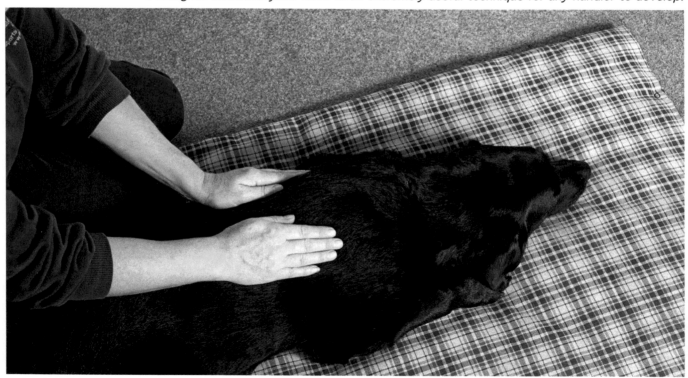

- Massage will have a soothing, calming effect on the animal that can be helpful in dealing with mouthing issues
- Massage can ease the discomfort that sometimes accompanies the rapid growth of bones and muscles in the first year of a puppy's life
- Massage can calm and focus an excitable, adolescent dog
- Massage assists the development of good spatial awareness in young dogs
- Massage reduces the risk of injury that can often accompany the boisterous activities that all puppies love: play, jumping, rolling and wrestling
- Regular massage is also an excellent way to establish and build a bond between you and your puppy

Recommended massage techniques for the puppy

- Gentle effleurage
- Passive touch
- Passive movement (only if the puppy is lying on his side and is completely relaxed)
- Gentle skin rolling
- Plucking
- Very gentle kneading
- Very gentle wringing

The application of each of the techniques recommended for puppies is described in detail in the chapters entitled *Massage techniques* and *How to massage your dog*. Remember, with puppies, to gear any massage sessions to their body clocks; perhaps wait until they are sleepy before you apply massage.

There are also specific instances when it is not advisable to massage your puppy.

- Do not apply massage if your puppy is unusually quiet, lethargic, subdued or withdrawn; if he has recently vomited or had diarrhoea; or if he is off his food
- Do not apply massage straight after or prior to the puppy eating (see contraindications in the chapter entitled *How does massage work?*)
- It is best to massage for a maximum of 30 minutes, and always end on a good note

Puppy exercises

There are some simple exercises and techniques that will help your puppy's general body

HEAT IN YOUR PUPPY OR OLDER DOG (ALSO APPLICABLE FOR MATURE AND VETERAN DOGS)		
HEAT DETECTION	**POSSIBLE CAUSE**	**COURSE OF ACTION**
Constant heat in a specific area or areas	A chronic or continuing problem possibly causing ongoing inflammation	Go to your vet for advice. Follow up with professional massage and possibly specific remedial exercises
Heat in a specific area after exercise	An acute and specific problem that may be caused by a recent accident	Go to your vet for advice. If appropriate, follow up with professional massage

Only carry out passive movement on a puppy when he is lying on his side and is completely relaxed. Care must be taken not to overextend limbs: stay well within your puppy's limits.

development; These exercises are also suitable for mature dogs.

- When taking your puppy for a walk, allow him time to amble, sniff around, and run at his own pace for a while. Also try changing pace sufficiently to break your puppy's rhythm but not overextend him: this will enable him to alter his head and neck position, which will exercise different muscles and allow him to move naturally at a comfortable gait

- Encourage young pups to walk on and over a small obstacle like poles (allowing him to walk with their head down), and then progress to perhaps getting him to sit on a small log. Helping spatial awareness and balance, by walking over obstacles with

Encourage young pups to walk on and over a small obstacle such as poles, as this older dog demonstrates.

the head down, will encourage hindquarter engagement

- Encourage your puppy to move very gently sideways; this can then be further developed by asking himt to walk sideways up onto a very low kerb, encouraging a sideways lift and step

Do not start any of these exercises until your puppy is at least six months old.

The mature dog

Like us, dogs can suffer from repetitive stresses, many of which are due to our environment, and which our dogs share. Although dogs seem to accommodate our world remarkably well, it is not unusual to find that, by the time a dog is six or seven years old, some of these stresses can begin to impact on his health. We are conditioned to believe that middle age for a dog starts at around seven years, but this is not true for many.

In fact, a dog thought to be suffering from the problems of middle or old age, or even osteoarthritis, may often be simply exhibiting signs associated with muscular stiffness and tightening, which themselves cause other allied issues. A repetitive strain issue will have a slow and insidious development, with small and almost insignificant changes evolving over a long period of time.

If ever your dog suffers an injury or accident there will always be an ensuing issue. Soft tissue problems underline a famous law of physics: "For every action there is an equal and opposite reaction." Muscle injury always needs specialist treatment or therapy: even if there seems to be no apparent lameness or problem, there will be change within the physiology of the muscle that will create further changes and imbalance, with resultant compensatory issues.

Many features that can cause repetitive strain issues or minor injury can be avoided or managed; it's just a case of perhaps managing

Signs of a possible muscular problem

- Lameness
- Behavioural changes: reluctance to play; quieter than usual; seeking dark areas; preferring to lie with back against a wall
- Just not right!
- Not bearing weight on one or more legs
- Reluctance to jump into the car, go down stairs, and so on
- Heat felt over one or more areas, either before or after exercise
- Uneven gait – hopping or pacing (left and right legs moving together)
- Coat deviations – not following a straight line (uneven waviness), or excessive curliness over a particular area (in heavily coated dogs this will be very difficult to identify)
- Stiffness after a walk or training – even if this does wear off
- Inability to settle when in bed
- Seeking something to rest the head on
- Collar fit becoming tighter (with no apparent weight gain)
- A change of shape (appearing larger over the forelimbs than the hindlimbs)
- Lameness developing a week or even months after an incident or accident
- A loss of muscle within the 'engine' hind legs
- Gaining of muscle tone in the front legs
- Offering a particular area of the body to you when you are sitting down
- Reluctance for certain areas to be groomed
- Excessive licking or chewing of an area (such as legs, feet or over hips)
- A dry nose in an otherwise healthy dog
- Tail carriage change (for example, carried to one side, or first horizontal and then dropping)

your dog's life slightly differently or simply being aware of areas that can cause problems. For instance, using a ramp to allow your dog to get in and out of the car is a good way of greatly reducing the stresses this causes; for a smaller dog the best option is to lift him in and out of the car. Introducing a fundamental change such as this will help ease stresses and assist in the maintenance of your dog's good shoulder and neck integrity.

One of the games or exercises enjoyed by most domestic dogs is ball chasing. For a domesticated dog, chasing a ball replicates the rabbit or small quarry hunting done by adult dogs in the wild. In

One of the major repetitive strain issues or minor injuries is caused through dogs jumping out of cars. These can be avoided or managed by using a ramp.

this natural situation, a dog would probably hunt three or four times a day in his pursuit of prey, and this meal would then last all day. The dog's body is constructed to twist and turn as it courses. The shoulders are built to absorb these rotational forces, but are not designed to do so continuously.

The problem is that most dogs love chasing a ball or similar object such as a frisbee, and will not stop until you do! It isn't necessary to completely stop throwing a ball for your dog to chase, but you should ensure he has warmed up first, and that you limit the game to four or five throws.

Even when your dog has matured and has settled into a good exercise and domestic routine, there are still some important things you can do to enhance his longevity and, more significantly, quality of life, by maintaining good, pain-free mobility.

Just like puppies, your mature dog should enjoy a variety of different types of exercise, though it's important not to create any habits that could impact on his muscular balance. Therefore, avoid always walking your dog on the same side when on a lead. You can provide a few challenges by having him walk over safe obstacles rather than always on the flat (this also encourages spatial awareness and engagement of the back legs). Perhaps also encourage him to walk sideways up a low kerb to stimulate the sideways action of the legs. Scenting exercises – finding objects around the house or in the garden – are fun activities that encourage extension of the neck and give the back a good, natural stretch.

Recommended massage techniques for the mature dog

Ideal techniques for dogs of this age are:
* Effleurage (generic, pre-event and post-event)
* Passive touch
* Passive movement
* Gentle skin rolling
* Plucking

continued page 90

FACTORS THAT CAN CREATE REPETITIVE STRESS ISSUES

Overexertion when cold

Always allow your dog to warm-up before exercise, and then warm-down again after exercise. Put your dog on a lead, or have him walk to heel for ten minutes before you allow him to run free. Put your dog on a lead ten minutes before you return from a walk to allow him cool down and stop panting

Overexercise at weekends

Dogs really suffer if they are over-exercised when puppies, but also if they are not fit. A typical scenario for many dogs is a quick, ten minute walk during the week, and then a five-mile hike at weekends. This can create minor fibre tears in the muscle that, on a long-term basis, gradually reduces muscle function

Too much ball/frisbee chasing

As explained in the text, too much twisting and abrupt stopping can cause shoulder and hip injuries through overstressing

Excessive jumping down from the car, high furniture, stairs, and so on

Jumping down requires a massive amount of concussion absorption through the neck and shoulders, and also through the vertebrae. The dog is built to jump down but not to do so repeatedly – and especially not from a car when its muscles are cold (at the start of a walk, say)

Repetitive exercise routine

If a dog is always walked the same way around an area, always on the same side when on the lead, and always allowed to run in one direction, he will become one-sided and unbalanced. Alternate the lead-walking side, and go the other way on a walk

Another major cause of repetitive strain in a small dog can be looking up constantly for treats when being trained, as the neck is elevated and held in that position for long periods of time

A competition in which dog and human are harnessed together for long-distance runs against the clock

A Canicross trailrunner dog being given a pre-event warm-up.*

- Kneading
- Wringing

(For instructions, see the chapters entitled *Massage techniques* and *How to massage your dog*.)

Massage for working and performance dogs

Working dogs, along with active companion dogs, can benefit hugely from massage incorporated into their exercise routines. This can be especially important for working dogs, since by their very nature they are often idle one moment and going flat out the next. Massage, with its warming properties, can assist in this dramatic change of tempo by preparing the dog for exercise – whether this is working or going on a regular walk.

The need for dogs to be correctly warmed is greatly misunderstood, and often totally ignored. A top athlete would never arrive at a sports event and then immediately perform: he or she wouldn't think of putting their muscles, tendons, ligaments and joints through exertion without first warming the tissues. Similarly, a rider would never take a horse from the stable and gallop from the yard without warming the animal first. So why is it that we do not naturally adopt this practice when we exercise, perform, or work our dogs?

Warming up before exercise aids oxygen and nutrient delivery within muscle fibres through enhanced arterial flow, extending performance potential. Warming – with progressive exertion – enhances the elastic qualities of the muscles, tendons, ligaments, and other soft tissue comprising the musculoskeletal system, and thus aids preparation for work and reduces the risk of injury.

Suggested dynamic warm-up

Initially work in straight lines, starting with a slow lopping jog for about 66ft (20m), then an active walk for 66ft (20m), a slow lopping jog for 66ft (20m), an active extended trot for 66ft (20m), a slow but active walk for 66ft (20m), an active extended trot for 66ft (20m), then a slow but active walk for 66ft (20m). Then exercise moving in 'letter' formation (for example, the letter 'M' or 'T'), ensuring your dog is walking on both sides of the handler.

Suggested dynamic warm-down

This should be an active, comfortable jog for three minutes that your dog finds agreeable, and covers the ground with ease, an active but comfortable walk for two minutes, and then a slow lopping jog for two minutes. Exercise using gentle and exaggerated 'letters' (as described above) to encourage gentle flexion through the back and neck. (The warm-down must not be too exertive: it is not intended to increase heart rate but to aid a resumption of a comfortable pumping action.)

Thermal images of a Border Collie before and after massage following light exercise. The red areas indicate the warmest tissue temperatures.

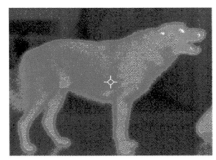

After light exercise and before a 90-second energetic massage.

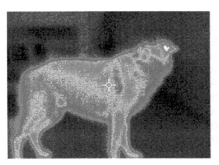

After light exercise and 40 minutes after a pre-event massage – and displaying considerably warmer areas over important muscle groups.

Pre-event massage

Massage can assist before or after a dynamic or active warm-up by targeting muscles and muscle groups, and by influencing deep vascular changes within the musculature. By applying a basic, yet energetic massage, the subsequent promotion of arterial blood flow to the muscle tissues will further aid the warming and nourishment of individual cells.

Adding passive movement to the routine can further aid preparation through the neural responses within the muscles and tendons. Muscles have an intrinsic 'failsafe' system to help prevent over-extension, flexion, adduction (limbs being forced towards the body), and abduction (limbs being pulled away from the body). This mechanism is within the muscle fibres and, especially, within the tendons. In an active dog, it is advantageous to prepare the muscles and tendons for expected movement. This will give neurological and muscular 'prior notice' to prepare for full range of movement, which will assist performance and also prepare muscles, further reducing the chance of injury.

Massage can also be used for the working dog when it is resting and likely to become cold between periods of activity. It can assist in maintaining muscle warmth between tasks to maintain steady thermoregulation and good muscle function.

Post-event massage

Warming down (cooling down) can also be assisted by gentle massage which will enhance the lymphatic system and venous return, helping to remove any metabolic waste that may remain within the fibres after the cessation of exercise. Also, and perhaps more importantly, gentle massage can assist the repair of microfibre tears within the muscles by the gentle enhancement of arterial flow to the area. These minor tears are thought to be the most common cause of post-

activity stiffness, and applying gentle massage will also assist corrective fibre alignment and gentle cleansing of toxins surrounding the area caused by cellular breakdown.

Passive movement also plays an important role here by gently easing the muscle fibres and joints to assist their realignment. The gentle movement can also enhance lymphatic and venous flow. This can be especially beneficial for the working dog which may not be given an opportunity to warm-down before being worked once more.

Pre-event/pre-walk massage technique

- Perform each of the strokes in effleurage three times over each area, and then do the same on the other side
- Use slightly more active rhythm than you would for a relaxing massage described previously, but ensure that you do not use extra pressure
- The massage is more effective if done 20 minutes before exercise

This massage technique is not intended to replace normal warm-up exercises. Perform all strokes a minimum of three times. The solid lines indicate the direction of movement. The broken line indicates movement performed on the inside part of the leg.

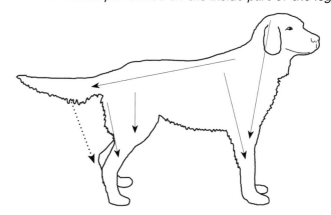

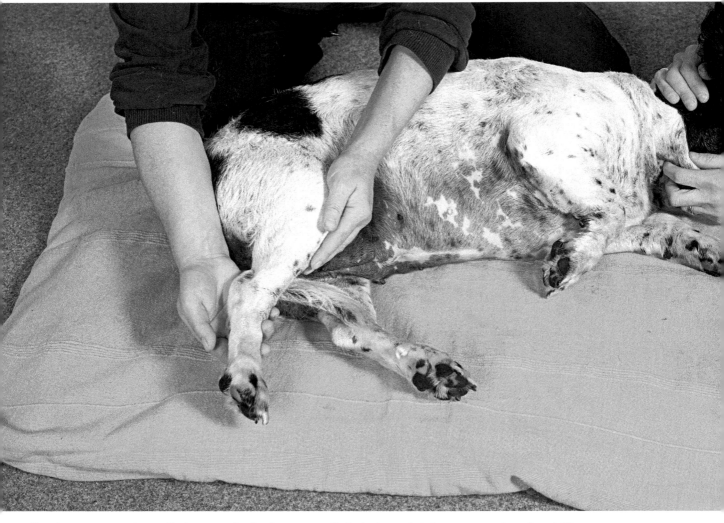

Passive movement, applied sympathetically, can really help to maintain movement and mobility.

- Effleurage is a 'stroking' technique and one that can be used with both pre-event and post-event massage
- It can be done with just one hand, or with one hand following the other. One hand must always be in constant contact with the dog

- It warms the area by friction and by increasing arterial circulation
- Keep your hands relaxed and follow the contours of the dog
- Remember to work with your dog and watch his reaction to gauge your 'touch'!

Post-event/post-walk massage technique

The same massage can be done for warm-down, but use a slow rhythm here, being aware of any heat or tension within the muscles or over the joints.

- It can be carried out up to four hours after exercise, when the body has regained homeostasis
- Again, it does not replace a warm-down, but enhances it
- It should be applied more slowly than pre-event massage to encourage relaxation
- When performing the massage, feel for heat and injury and manage appropriately
- It is intended to relax muscles and influence arterial blood to promote healing and flexibility
- It is intended to influence lymphatic and venous return – therefore, observe lymph node positions and stop at these points before progressing

The veteran dog

It's important that the veteran dog receives the right sort of exercise and muscle care, although we should not assume that a dog without underlying skeletal or neurological issues will necessarily become stiff and arthritic when he gets older: there are many ten- and eleven-year-old dogs competing at top levels in agility because they have received good muscular care.

If your dog is well constructed (in other words, his body is well balanced), the likelihood of issues in later life becomes less likely, especially if any problems or traumas earlier in the dog's life have been spotted, and the causes dealt with appropriately.

If a veteran dog is showing signs of discomfort, there are several ways to make life easier. For example, shorter, more frequent walks are better than one long one. Massage can also help an older, stiffer, less active dog. A more recumbent dog will not be stimulating his venous or lymphatic systems, so the percentage of venous blood that could be pooling in areas of restricted activity will be high, and natural healing potential reduced. Smaller and more regular walks will enhance the venous or lymphatic systems without fatiguing your dog.

Older, stiffer dogs tire more quickly because their muscles are not working in the correct patterns: some muscles are barely working and others are overworking. The overworked muscles will cause the dog to be fatigued. When muscles tire they are prone to injury (minor fibre tears) that will further reduce their integrity and slowly reduce mobility. This will add to the already established compensatory issues.

Gentle effleurage – the most versatile of massage techniques – will provide a relaxing and therapeutic massage to help with any muscular stiffness, and really assist the lymphatic system and venous blood flow.

Skin rolling can be especially helpful to the older dog, especially one who has been very active and has probably picked up a few injuries along the way. Skin rolling will help any superficial adhesions that can impede mobility – and is also very relaxing!

Another very helpful massage technique is passive movement, which can help to maintain good movement and mobility, and also influence good joint health. It must always be applied by following the rules and in conjunction with other techniques, but even multiple osteoarthritic joints can benefit greatly from the treatment.

7
MANAGING MUSCLE REPAIR, INCORPORATING MASSAGE

One of the most difficult aspects of canine care is knowing how to manage your dog's recovery after injury, after an operation, or other forms of treatment. Unfortunately, there are no hard and fast rules here, although an understanding of tissue repair and how the body recovers in different situations may help you organize an appropriate routine for your dog.

The most difficult question is to know when a dog needs to rest and when he should be exercised, at a critical time when so much damage or hindrance of good recovery can be inflicted. The main problem is that we cannot see inside our dog, and nor will most give us clearly identifiable signs that what we are doing is correct.

Before any course of recuperative action is taken, you must first consult your vet, or at least discuss your planned strategy with him or her. It's also vital to know when your dog needs to rest and when he should be exercised as part of a recovery programme, because getting this wrong can result in much damage, or, at the very least, can hinder good recovery. For example, your vet might diagnose a vertebral disc injury that could be exacerbated by exercise and massage, leading to serious consequences. In all of the following situations, exercise management should follow a discussion with, or examination by, your vet.

Quite often, dogs do not give clear signs that anything is wrong.

Muscle and soft tissue injury

Muscular injuries occur in a variety of ways. Typically, they can result from a dog catching his foot in a hole when out for a walk, during an over-exuberant play session, or falling or slipping off a wall or a style. Equally, the injury can be incurred when he is performing or working.

As well as injuries to muscles, soft tissue injury can include damage to tendons and ligaments: injuries that require a more specialized recovery plan. However, for first aid purposes, the course of action described here is appropriate and should be adopted.

As soon as possible after a suspected injury has occurred, run your hands over your dog to feel for heat in the same way as described before or during a massage (see pages 79 and 81, in the chapter entitled *Massage, exercise, and muscular health*). This is much easier to do if heat detection has become routine, because you will then know the difference between normal and abnormal heat levels in your dog.

After an injury, varying amounts of heat are quickly radiated, depending on severity and depth of injury. Note: if your dog is displaying profound lameness, and/or is unable to bear weight on the affected limb, veterinary attention must be sought immediately: this course of treatment is intended for minor injuries only.

Dealing with muscular injury

If you suspect a muscular injury, carry out the following procedures.

1   Restrict exercise immediately. If out walking, put your dog on a lead (or carry him if this is comfortable for both of you), and get home as soon as possible

2   Immediately place ice, or something cold, on the injured area. This is obviously not possible if you are away from home; if, however, you are at a show or working event,

then ice or something cold should always be carried in case of such an occurrence. Leave the ice in place for a maximum of five minutes and then remove it. If the injury affects a joint, leave the ice in place for just three minutes before removing. Ice must never be placed in direct contact with the skin; wrap it in a cloth, plastic bag, or something similar first

3   After removing the ice wait for a few minutes. If the heat returns, reapply the ice for the same amount of time

4   Continue reapplying ice and removing it until the heat has dissipated

5   Rest the dog for 24 hours. The body will be trying to heal the damaged cells by forming the equivalent of a scab over the damaged tissue. The way muscles heal is similar to the way that skin heals, and the tissues must not be disturbed as they attempt to repair themselves

6   Keep your dog calm for another two days. Provide quiet lead exercise (not on an extending lead), and do not allow him to overexert himself. In terms of exercise, little and often is better – ten to fifteen minute walks, two to three times a day on the flat to begin with. Try to prevent play, jumping on and off furniture, or in or out of the car

7   On the third day, perform gentle effleurage, following the lie of the coat. Do not use petrissage or passive movement at this stage

8   For the next two weeks, perform gentle massage and allow restricted exercise. It is important that your dog is exercised at this time because scar tissue is forming, and you must ensure that the fibres are healing without binding together to form a knot. Note, however, that too much exercise will cause further injury to the area

9 After two weeks, allow your dog off the lead for short spells while still trying to prevent excessive play, or jumping up and down Exercise can be increased in small increments

10 Gently introduce passive movement to regain full range of movement and ease the fibres into full functioning; perform skin rolling to ease any adhesions that may have formed within the skin

11 After four to six weeks your dog may be back to near fitness, although this timescale depends on the severity of the damage

12 For complete care, organize an appointment with a canine muscle therapist to ensure thorough and appropriate healing

The repair cycle

To understand further how muscles heal, the following is a breakdown of events that occur during and after injury, with approximate timescales.

1 Bleeding phase (bruising). When a muscular injury occurs, a large amount of blood will escape from the muscle fibres through tearing or crushing when the cells are damaged. Average timescale is six to eight hours

The importance of ice

When a muscle is damaged, the cells are fractured and the blood and other cell contents are spilt into the surrounding tissue spaces

The body instantly recognizes that these substances are no longer contained within a cell wall, and therefore treats them as toxins. All toxins detected by the body are managed as if they are dangerous, and the body's response is to send additional blood to the area to kill off these potentially harmful organisms (in other words, the cell contents). This causes an excessive amount of vascular (blood) activity in the area. The small vessels serving the area of injury become overloaded by the response, and, in turn, become damaged and also rupture

If ice is applied to the area, it has the effect of gently constricting the blood vessels and reducing blood flow, thereby managing to contain the injury and reducing additional damage. By creating this restriction, the injury site can be limited to the affected area only, without further damage from engorgement of surrounding tissues, which is why it's important to apply ice or something cold as soon as possible after injury

It is essential that ice is applied for five minutes or less only to control blood to the area; if it's applied for longer, the body will register that there is a dangerous restriction of blood to the area (a vasoconstriction), and will counteract the effects by delivering more blood, with the possible result of further damage

Note that ice is only appropriate in acute injury, or if your vet suggests its use post-operatively; it is not appropriate in isolation for chronic or ongoing heat or injuries

management, remodelling should resemble the original muscle

The recovery schedule described here may seem excessive – particularly if your dog gives the impression of having recovered – but if you pay attention to the healing processes and allow for proper recovery, the incidence of compensatory issues will be greatly reduced. Muscular injuries can be invisible and insidious, and an injury that is not properly healed can manifest itself at a later stage in the dog's life – sometimes years later.

Post-operative treatment

The following advice is of a general nature only; it is therefore important that you discuss the specific details of your dog with your vet.

Give your dog time to recover from injury by allowing only gentle exercise.

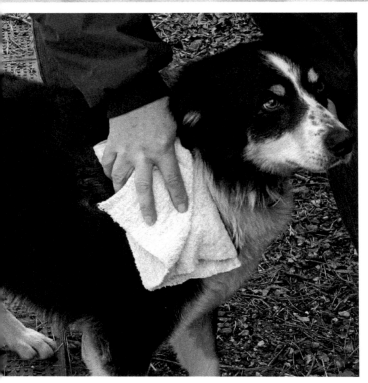

When applying ice, make sure it is never placed in direct contact with the skin.

2 Inflammatory phase (heat detection). This is an essential component of healing. It begins a few hours after the injury occurs, and increases to a maximum reaction in two to three days, although it can last much longer

3 Proliferation. This is the phase of scab – then scar – production. The onset is 24 to 48 hours after the injury, with a peak reached in two to three weeks

4 Remodelling. An essential phase of muscle repair, it starts within one to two weeks and continues for sometimes weeks afterwards until there is a functional scar. With good

MUSCLE FIRST AID CHECKLIST

Action	Timescale
Ice	Apply for a maximum of five minutes, remove then check after ten minutes. If injury is still hot, reapply ice
Rest	Total rest for 24 hours
Rest (restricted exercise)	Two days
Gentle effleurage	Three days after injury
Gentle exercise	For a further two weeks
Gentle exercise introducing off-lead activity and gradually building up to normal activity	For a further two weeks
Introduce passive movement and skin rolling	Two weeks after injury, when healed, the muscle fibres need to be taken through full range to regain elasticity, and as full a function as possible

By using this technique, inappropriate scarring and congestion of the muscles should be minimized

EMERGENCY OPERATION

The kind of massage you can apply in this situation depends totally on the type of operation or procedure that your dog has had. With any type of emergency, or if your dog has been seriously ill, you must follow to the letter any instructions from your vet.

If you are able to visit your dog in the hospital, passive touch and plucking can be extremely comforting to the animal, and also gives you something tangible to do. Obviously, this must be cleared by the nursing and veterinary staff first. Passive touch can help relaxation and give comfort to your dog, whether it is conscious or unconscious. Plucking (not around the site of the injury or wound) can help to gently stimulate your dog, especially if he is struggling with the effects of the anaesthetic. It, too, is relaxing, and does

not impose any physical pressure or weight on the body.

When your dog returns home, gentle effleurage will help the detoxification process that would otherwise be hampered due to the dog's lack of mobility. Only do this for five minutes at a time, so you that do not overstress the body.

PLANNED PROCEDURE

Prepare your dog for a 'routine' operation by using effleurage and other relaxing techniques before you leave him. (This will also help you, by giving you something to do!)

X-RAYS

If your dog has been X-rayed, the likelihood is that it will be for a mobility issue, and he will be stiff and sore when he gets home. This is because, to get the best picture for diagnostic purposes, he may have been arranged in awkward positions when anaesthetized; unfortunately, this is unavoidable. To ease this potential problem, effleurage your dog and, if feasible, apply passive movement before he is x-rayed to help the body prepare.

Do not, of course, apply massage if your dog has been X-rayed for one of the several conditions already mentioned where such treatment is not beneficial (chapter 3, page 41).

When your dog returns home, gentle effleurage can help the venous return remove the anaesthetic from his body more quickly. Do not massage for more than ten minutes at a time, leaving 12 hours between treatments to enable the body to dispose of the toxins without overloading the system. If your dog is very sleepy and struggling with the effects of the anaesthetic, plucking

Plucking is a gentle, stimulatory technique that is not manipulative, and can be extremely comforting and relaxing.

can help by gently stimulating awareness, while maintaining a relaxed demeanour.

The following day, effleurage – and, if appropriate to the condition, passive movement also – can further assist venous return.

Occasionally, when a dog returns from an X-ray, movement is looser and apparently freer. This can be because the position your dog was placed in to be X-rayed – for example, with hind legs stretched – has assisted muscle length, so the issue could be muscular, or at least have muscular implication. Therefore, you should perhaps consider muscle therapy for your dog, along with any treatment recommended by your vet.

NEUTERING

This procedure will result in a wound that requires immediate healing. Ask your vet about mechanical massage techniques before applying massage, since these may influence unnecessary blood flow to the area, and could therefore be a contraindication.

Plucking is an appropriate technique if your dog is content with the procedure; followed the next day by skin rolling. When the stitches have been removed from a spayed bitch, or the wound has healed, use effleurage and passive movement to re-establish mobility within the pelvic region.

LIGAMENT AND TENDON OPERATIONS

With these operations, it is vital that you seek post-operative advice from your vet. If possible, you should also seek further guidance and treatment (working in conjunction with your vet) from a myotherapist or qualified animal physiotherapist. Ligament and tendon operations generally require a strict protocol of post-operative treatment and exercises that will aid appropriate healing. Using the advice given, the exercises should generally begin as soon as possible after your dog has recovered from the effects of the operation.

A good myotherapist or physiotherapist will offer advice and suggest exercises to be done by the

continued page 106

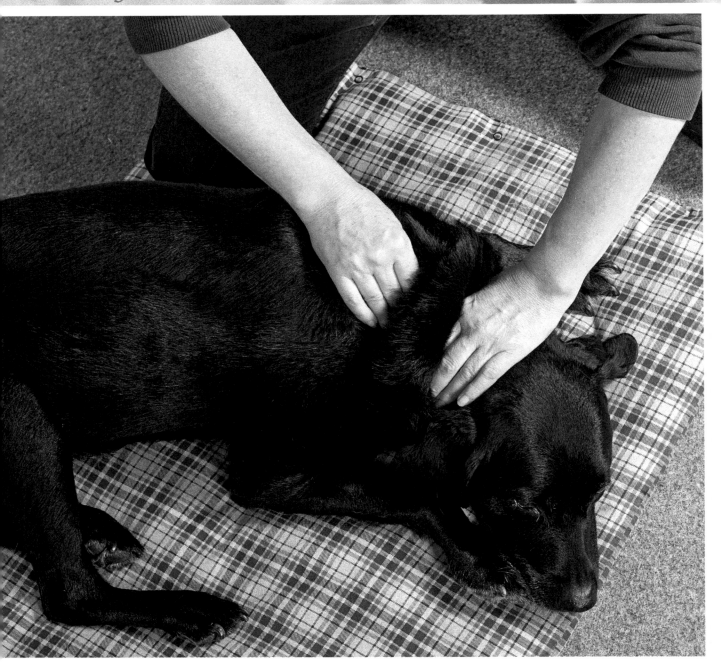

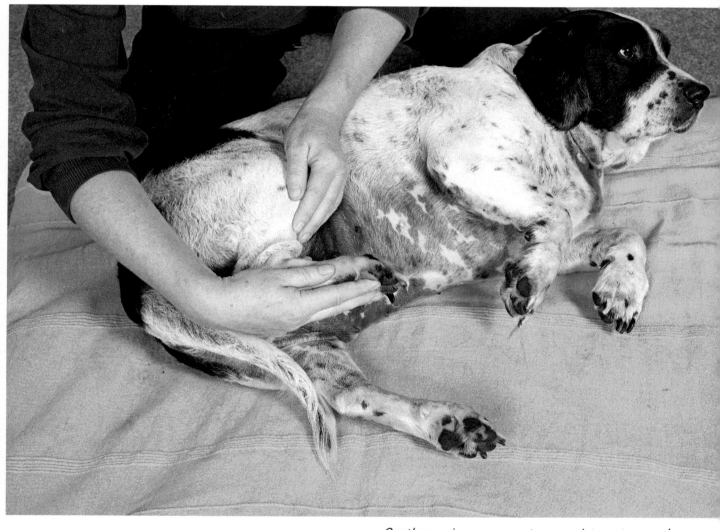

Gentle passive movement can assist post-procedure recovery, but should only be carried out with veterinary approval.

Oppposite: Gentle effleurage can assist with disposal of toxins following an anaesthetic.

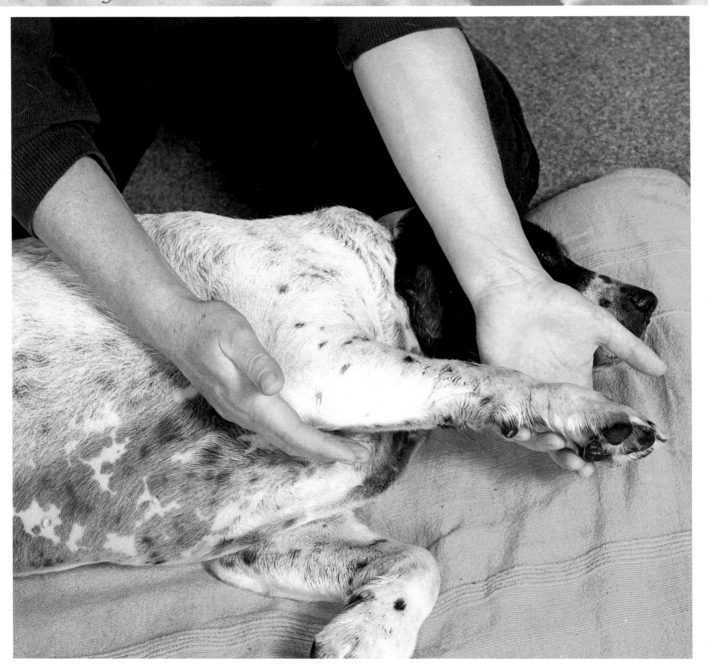

Use effleurage to re-establish mobility.

Opposite: Take great care when applying passive movement to stiff dogs, or those requiring remedial therapy.

owner at home that will encourage appropriate healing, and a speedier and more robust return to normal (see the chapter entitled *Other techniques*)

Post-treatment exercise

It's very important to restrain your dog for at least two days following any kind of physical therapy (massage from a therapist). After appropriate therapy, changes will be elicited, including those meant to break down adhesion and scar tissue. It's therefore important to allow the body to reheal without overexertion, so that necessary changes and muscle patterning can occur.

A good therapist will provide recommendations and instructions according to the treatment given, and the age and condition of your dog. When a dog is nearing the end of his treatment plan, there will be a need to extend and flex his joints to re-establish better function (see also *Scenting exercises* in this chapter), and this is the time when he may need some hill exercise.

Also of benefit could be hydrotherapy, to encourage resistive exercise that will help strengthen and develop muscle tone without exerting undue pressure on the body.

LEAD EXERCISE
Lead walking is a vital method of reintroducing or restricting exercise for your dog. It is probably one of the least understood procedures, however, and is not always used effectively as a result.

Lead exercise is usually best done using a short lead so that your dog is restrained, and stays close to you. It is inappropriate to use an extending or 'flexie' lead in these circumstances, because this does not provide the restraint needed: letting the dog run and then making him come to a sudden stop could actually hinder healing.

Make sure you exercise your dog on both sides of you; not just the side to which you are both accustomed.

POST-TREATMENT
Used to restrict a lively dog so that he will settle and not be overexerted, thus allowing changes that have been brought about by the treatment to take effect. Controlled movement allows the muscles to work in a more appropriate way, encouraging a healing process. This is generally given a timescale according to the dog's age, condition, normal activity levels, and treatment; it also means that your dog works in straight lines only, without twisting and rotating.

Post-operative exercise

Depending on the operation, walking is important to mobilize your dog, help the healing process, and begin to regain a good range of movement (see also the chapter entitled *Other techniques*).

The type of surface used for walking should be flat and smooth, because any form of irregularity will exacerbate a problem. Be aware of your dog's gait or pace, so that during rehabilitation you are adapting to his rhythm rather than he to yours. When your dog is improving, alter your gait to encourage a variety in gait pattern: this will gently encourage his use of different muscles, and help re-establish good muscle patterning.

Scenting exercises

One of the most under-exploited remedial exercises for a dog is encouraging him to use his nose. Scenting exercises within the home or garden can help stimulate him, but in a constrained way that will provide mental exercise but not physically overexert your dog. Tiny morsels of food or toys can be placed in safe locations (which do not involve climbing or running) for him to find. You can even buy specially-designed toys for the purpose. Remember that a dog's scenting ability far exceeds our own, so each time extend the challenge by leaving progressively smaller items for him to find, and then reward success with a larger

Lead walking is a vital method of reintroducing or restricting exercise.

Scenting exercises can help occupy the recovering dog by stimulating the brain while not over-exercising the body. This specially designed device encourages the dog to find hidden treats.

piece. These games can extend over a greater distance when your dog is ready to increase his mobility.

Scenting games have many advantages:

- They are fun and stimulating for your dog
- They encourage him to drop his head, thus helping stretch the back and neck
- A head-down position promotes better hind leg movement, especially beneficial for cruciate (hind leg ligament) problems and operations
- It gives your dog an objective when normal walking has to be restricted
- It gives your dog one-to-one time with you

Scenting exercises encourage the dog to drop his head, thus helping to stretch the back and neck.

8
OTHER TECHNIQUES

Until quite recently, some of the therapeutic treatments now widely used on animals were not even readily available for humans. Today, however, 'total health care' seems as appropriate for our pets as it does for us. With so many techniques and treatments available, however, one of the difficulties faced by both vets and dog owners is knowing which type of therapy is most suitable for a particular condition, and, most importantly, which protocol or procedure is appropriate for it. To further compound this dliemma, there are no rules for a particular pathology (condition) or injury, and each should be considered and treated on a case-by-case basis.

Whatever therapy you feel you would like to explore, always discuss this course of action with your vet first (see the chapter entitled *Using your vet*). This should be a two-way discussion, with both parties receptive to the other's opinions and suggestions. To aid these discussions, it's advantageous for there to be a degree of understanding about what is involved in different therapies: it's useful if you have at least some knowledge of a particular treatment so that you can ask appropriate questions – both of your vet and also of the suggested practitioner or therapist – before you begin any treatment programme

Balance and mobility

In most cases of rehabilitation, irrespective of whether the problems are of a degenerative or post-operative nature, your aim is to achieve balance and mobility, as the direction of the following text indicates, and demonstrates why it is important to follow a system of combined therapy and exercise in the correct order.

Why the establishment of balance is the key to improved mobility:

BALANCE IS TO ESTABLISH
Both sides of your dog are equal
The hind legs are the 'engine'

Your dog is moving from his hind legs; he does not have 'front-wheel' drive

Once balance has been established, mobility can be encouraged.

MOBILITY IS TO ESTABLISH
Improved mobility within muscles and muscle groups
Muscles are able to contract and relax as fully as possible to demonstrate improved range of movement and then improved muscle tone

Once mobility has been improved
Individual cells are mobile and able to take up nutrition and oxygen

Mobility within joint
Range of movement and joint health will be enhanced through good balanced mobility. This applies to both vertebral and limb joints

Addressing the issue of balance first will help you to achieve the ultimate goal of enhanced or normal mobility.

If a dog has a chronic injury caused by a degenerative disorder, no amount of therapy will totally recover his balance. However, by maximizing balance you will aid his mobility, with the dog moving more properly (correct muscle patterning), easing stress on overused muscles and joints caused through previous imbalance, and also ease subsequent pain perception, therefore further assisting mobility.

Regaining or redressing the balance will help an injured dog resume his correct movement pattern, which is important to do before mobility or strengthening therapy.

If your dog has been through some form of operation or orthopaedic procedure, the likelihood is that he was lame and limping for some time before the operation, which would, in itself, have caused imbalance. Post-operative therapy should

A dog with a severe muscular problem in the neck; unable to lift his head into a natural position.

therapist or myotherapist may use many different techniques to the ones described in this book to effect deeper and more profound change within your dog's musculature. If you decide to use a therapy of this type, however, always ensure that the therapist is fully accredited and insured, and that they gain veterinary consent before treating your dog. (see the chapter entitled *Using your vet*).

There are many different kinds of massage therapist. And, as in many professional bodies, there are various levels of training and practice. Some therapists perform a set choreograph or predesignated routine only, whilst others will use highly refined palpation skills to target the areas of

The same dog, three weeks later, much improved and with good neck movement.

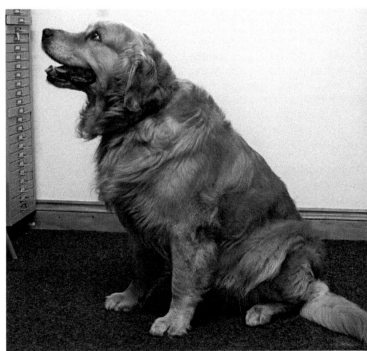

therefore be aimed initially at addressing the balance issue, which, in turn, will assist resumption of correct or improved mobility.

'Balancing' therapies

Therapies that will encourage muscular balance are:

- Myotherapy
- Massage
- Physiotherapy
- Acupuncture
- Gentle osteopathy/chiropractic

Myotherapy (massage)

As already mentioned, this is not a definitive guide to massage techniques. A professional massage muscular issue. Using such techniques, they should ultimately be able to identify 'cause and effect' – in other words, it should be possible for them to not only locate the source of the main issue and identify compensatory issues, but also treat them accordingly.

Acupuncture

The treatment of ailments using acupuncture was developed in China over 3000 years ago, and this is now used to treat a diverse array of medical conditions in the west. Acupuncture involves stimulating nerve fibres, which brings about a change in the muscles, nerves and cells of the body. In the UK, it is popularly used in animals for the management of pain in arthritic conditions.

'Strengthening' therapies

HYDROTHERAPY

As well as being an efficacious therapy, for many dogs this is also an extremely enjoyable one; therefore, it can really help a dog physiologically and psychologically. It works primarily by providing 'mobility' without the need for the patient to endure the hard impact experienced when walking or running. By eliminating pounding through the limbs and into the back, pain perception can be reduced. This encourages the dog to swim more, thus exercising the muscles and mobilizing the joints. It can also help ligament and tendon injury repair through resistive exercise – in other words, using the dog's limbs to push against the resistance of the water as he swims.

There are many types of pools or water walkers available for this treatment, although the applications can vary. Whatever hydrotherapy option you choose, try to establish a degree of muscular balance in your dog beforehand, so that when he does become more mobile he is better able to maximize movement and gain positive benefit.

CASE STUDY: MARLEY

Marley originally received acupuncture to help him when he began to limp on one of his front legs. This resolved well, and the acupuncture treatments were continued to help Marley maintain his overall mobility. The treatments stimulate his muscles and help to combat the weakness that is occurring as part of his aging process. He is keen and still able to go for walks at over nineteen years of age!

Marley undergoing treatment.
(Courtesy Suzannah Stacey BSc BVM&S MRCVS Cert Vet Acu (ABVA))

EXERCISE PHYSIOLOGY

This is an excellent method of supporting balance therapies by devising exercises specific to your dog's needs. These exercises will help re-establish correct muscle patterning and, equally importantly, develop good core strength and stability.

Canine core stability and balance is an extremely important therapy using exercises that incorporate movements and actions that dogs

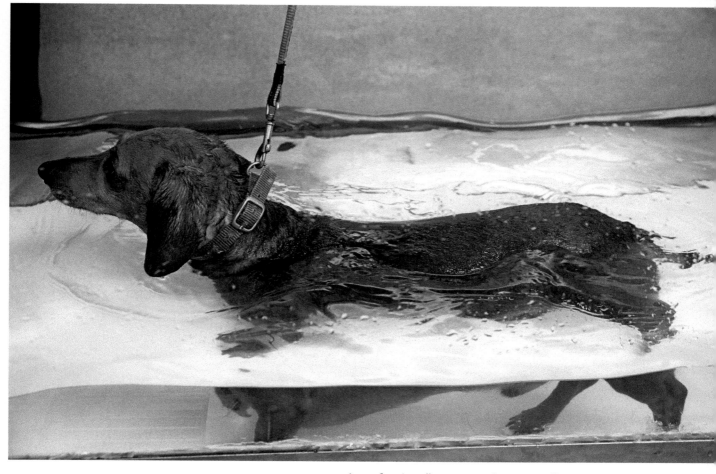

A professionally-managed water walker can assist recovery.

normally include within their natural routine, but very slowly. However, it should only be applied after being taught how and when to use it by experienced Canine Exercise Physiologists.

Other techniques

Currently, there are many books and DVDs available advocating different forms of stretching,

tricks and 'balance' exercises. If these are not applied correctly, they have the potential to cause serious problems and exacerbate existing injuries or conditions through lack of good and appropriate application.

Only a trained professional should apply stretching techniques to your dog, because much harm can be done if this very extreme form of canine massage is used inappropriately. When we stretch, we simply stop before we reach the point

Free swimming can assist mobility, and give a feeling of freedom for the incapacitated dog.

at which we might strain ourselves; there are also exercises designed to stretch a horse's muscles, but, as with humans, it's difficult to overstretch them against the horse's will.

However, it is very easy to unintentionally overstretch a dog's muscles, which can lead to serious problems as well as inflict pain. It is important, therefore, to make the distinction between passive movement and passive stretching.

During the massage technique known as passive movement (see the chapter entitled *Massage techniques*), the operator takes the joint *through* its natural range of movement, whereas a stretch takes a joint *beyond* its natural range of movement, 'overriding' the natural stop in the dog's range of movement.

There are also articles and information about putting dogs on balance balls (Swiss balls used in human pilates), and other such apparatus. These 'tools' were designed for highly skilled practitioners to use in human core stability classes and exercises. Exercises like these can, in my opinion, cause damage to your dog, and their use does not resemble anything they would do as natural behaviour.

Canine core stability and balance is an extremely important therapy using exercises that incorporate movements and actions which dogs naturally include within their normal routine. However, it should only be applied after being taught how and when to use it by experienced therapists and trainers.

Opposite: It is important to assist your dog's flotation if there is a mobility or neck issue.

Slowly walking over irregular poles, encouraging the support and lift of all four legs independently.

GLOSSARY OF TERMS

Words in blue refer to other glossary entries

Acupuncture
A treatment aimed at restoring the balance of the universal energy Qi (pronounced 'chee') within the body, through the painless insertion of fine needles into specific points on the body

Animal protection laws
Laws designed to protect dogs and other animals from inappropriate therapy by unqualified people. They comprise the Veterinary Surgeons Act 1966 (also sometimes called the Veterinary Act), and the Animal Welfare Act 2006; in Scotland, the Animal Health and Welfare Act (Scotland) 2006

Appendicular skeleton
The parts of the skeletal system comprising the fore- and hindlimbs, including the shoulder and pelvis, that attach to the axial skeleton

Axial skeleton
The parts of the skeletal system comprising the skull, bones of the ear and throat, the ribs and the vertebrae

Autonomic nervous system
The non-conscious part of the peripheral nervous system that control activities such as respiration rate, heart rate, heat regulation, digestion and urination. It is divided into two parts: the sympathetic nervous system (controlling 'fight and flight'), and the parasympathetic nervous system (controlling 'rest and digest')

Biomechanics
The study of mechanical principles as they relate to living organisms – for example, the system of levers and muscles that operate the limbs

Central nervous system
The part of the nervous system that comprises the brain and the main nerves running down the spinal cord. See also peripheral nervous system

Connective tissue
A type of tissue within the body responsible for binding and supporting other internal structures

Constriction (of vessels)
The narrowing of a blood vessel resulting from contraction of the muscular wall of the vessel. Also called vasoconstriction. See also dilation (of vessels)

Contraindication
A condition or factor that increases the risk of a surgical or therapeutic procedure

Cutaneous trunci
A thin sheet of muscle that covers the whole of the chest and abdominal regions of the dog; it lies directly under the skin

Deep muscles
Muscles within the body lying beneath superficial muscles

Dilation (of vessels)
The widening of a blood vessel resulting from relaxation of the muscular wall of the vessel. Also called vasodilation. See also constriction (of vessels)

Distal
See directional terms

Dorsal
See directional terms

Dynamic or active warm-up
Warming up the body muscles by using exercises such as walking and running

Engorgement
With regard to a muscle, a condition in which the tissue is damaged and thickened, possibly due to a

swelling/excessive fluid, and is therefore unable to function fully

Fascia
The soft tissue component of the connective tissue system. It penetrates and surrounds muscles, bones, organs, nerves, blood vessels, and other structures. Fascia is a continuous, three-dimensional web of tissue that extends throughout the body, from just under the skin to deep within the interior. It helps maintain skeletal and muscular form, provides protection, and acts as a shock absorber for the body as a whole

Handler
Anyone who is in control of an animal such as a dog

Homeostasis
The body's regulatory system for its internal environment. It maintains a stable and constant condition, irrespective of external situations

Hydrotherapy
A therapy that involves totally or partially immersing in water an animal such as a dog to aid fitness or rehabilitation

Inflammation
The intricate biological response of tissues to harmful stimuli, such as damaged cells, irritants or infection. It is a defensive move by the body to combat toxins and promote tissue healing

Innervate
To stimulate a region of the body with a nerve impulse. See also muscle firing

Involuntary muscle
Muscles not under conscious control which work without our awareness; for example, heart muscle. See also voluntary muscle

Massage
The manipulation of soft tissue structures of the

body. It has both mechanical and reflex effects on the body. See also mechanical response and reflexive response

Mechanical action
The movement (response) of body tissues as a result of massage or other forms of therapy

Metabolism
The chemical reactions that take place in a living organism in order to maintain life. The body's combined physical and chemical processes

Motor nerve
A nerve originating in the central nervous system that carries nerve impulses from the brain and spinal cord through the peripheral nervous system to directly or indirectly control the muscles or glands. Also called motor neuron or efferent nerve. See also sensory nerve

Muscle
Specialized body tissue with the ability to contract and relax, and thereby produce movement

Muscle patterning
Muscles working in the correct order and 'drive' to provide either stability or movement, according to their function

Muscle firing
The response that occurs in a muscle as a result of an innervation

Myotherapy
Muscle therapy as used in Galen Myotherapy®

Neural pathway development
The stimulation of nerves through appropriate exercise or massage

Nociceptor
One of the damage-limitation neural (nerve) receptors in the body that receives and transmits pain signals

Palpation
Physical examination by the hands of the practitioner to determine and assess normalities and abnormalities

Passive movement
Taking the limb of an animal (such as a dog) through its 'range of movement,' or as far as it can comfortably go without resisting. This helps to adjust muscle memory and can improve mobility

Passive stretch
A form of therapy in which the limbs are taken through a movement beyond their normal range. Only properly trained persons should carry out this therapy

Pelvic limb
A term for the hind leg or limb

Peripheral nervous system
The part of the nervous system that exits from the vertebrae and serves the body's systems and organs. See also central nervous system

Physiological
Relating to the functions of living systems

Plane of movement
Term used to describe how the parts of the body move in relation to the axis

Proximal
See directional terms

Psychological
Relating to mental functions, behaviours and perceptions

Range of movement
The degree to which a limb can flex (close the angle) and extend (open the angle). This is dependent on how the animal is constructed, and the integrity of the soft tissues and participating joints

Reflexive action
An action that works indirectly on the nervous system by triggering the sensory nerves and also hormones through the application of different massage techniques and other forms of therapy

Sebaceous gland
A gland in the skin, usually associated with a hair follicle, that secretes oil for an animal's coat

Sensory nerve
A nerve that carries impulses from the receptors (those that sense pain, heat, etc), back to the central nervous system. Also called efferent nerve. See also motor nerve

Skeletal system
The rigid framework of bones inside the body of animals such as fish, reptiles, birds and mammals (including dogs)

Spasm
A sudden and involuntary contraction of a muscle or a group of muscles, usually to protect a damaged or vulnerable joint

Spatial awareness
The ability of an animal to be aware of its position in space and the location of objects in relation to its own body

Superficial muscles
Muscles within the body lying over deep muscles

Thermoregulatory system
The internal heating or cooling mechanism that keeps the body temperature of some animals within certain boundaries, irrespective of external conditions

Thoracic limb
A term for the front leg or forelimb

Ventral
See directional terms

Voluntary muscle
Muscle under conscious control. Also known as skeletal muscle. See also involuntary muscle

DIRECTIONAL TERMS
Dorsal
Towards the upper surface or towards the back; also the upper portion of structures, including the front surface of the carpus and tarsus bones

Ventral
Towards the lower surface or towards the belly; also the lower portion of structures. (However, in relation to limbs, then either palmar (referring to the foreleg) or plantar (referring to the hind leg) is used instead)

Proximal
The part nearest to the main mass of a body or organ

Distal
The part furthest from the main mass of a body or organ

Cranial
Towards or near the head; also, on the limbs, proximal to the carpus bones

Caudal
Towards or near the tail: also, on the limbs, proximal to the palmar and planter aspects

Plantar
The pelvic limb paw on which the pads are located

Palmar
The thoracic limb paw on which the pads are located

PLANES OF MOVEMENT
Frontal
Moving the limbs away from or towards the midline of the body

Transverse
Moving the limbs in a twisting movement

Sagittal
Moving the limbs forwards and backwards

OLDER DOG? NO WORRIES!

Maintaining physical, mental and emotional wellbeing in your golden oldie

Sian Ryan

Hubble & Hattie

Drawing upon the latest research to provide ideas for maximising your dog's well-being as he ages, the individual chapters allow you to develop your own care plan for your dog, to incorporate new or amended ideas into your daily routine, and to make simple changes to your home, garden, car, and walks, to ensure your older dog is happy, safe, and invigorated.

HH5366 • paperback • 20.5x20.5cm • £13.99* • 96 pages • 100 colour images • ISBN: 9781787113664

Communicating with and understanding our dogs establishes trusting relationships, and relieves frustraiton and improves lives at both ends of the leash.

This innovative book explores the juncture between hearts and minds, where true understanding begins.

HH5305 • paperback • 15.2x22.5cm • £10.99* • 96 pages • 15 colour images • ISBN: 9781787113053

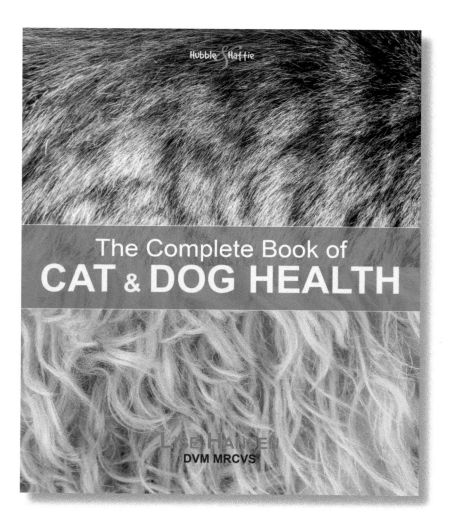

The Complete Book of
CAT & DOG HEALTH

Hubble & Hattie

LISE HANSEN
DVM MRCVS

Written by a qualified
veterinarian, this is
a complete and
comprehensive guide
to health care for cats
and dogs, providing
invaluable advice on
essential aspects of
care, such as diet and
vaccinations, as well
as a guide to holistic
treatments

HH5415 • paperback • 20x24cm • £19.99* • 234 pages
•58 colour images • ISBN: 9781787114159

For more info on Hubble and Hattie books, visit our website www.hubbleandhattie.com
email info@hubbleandhattie.com • tel 44 (0) 1305 260068 * prices subject to change •p&p extra

Index

RECORD SHEET

Your dog's name ....

Date ....

Comments ....
....
....

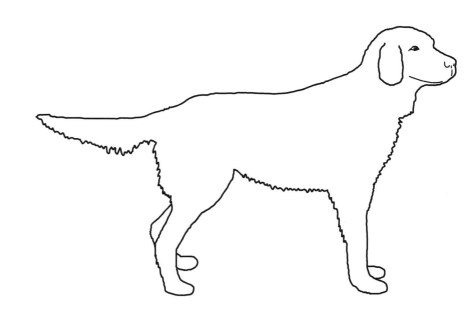

Key
*   heat
x   tension
+   sensitivity
√   improvement

The Hubble & Hattie imprint was launched in 2009 and is named in memory of two very special Westies owned by Veloce's proprietors. Since the first book, many more have been added to the list, all with the same underlying objective: to be of real benefit to the species they cover, at the same time promoting compassion, understanding and respect between all animals (including human ones!)

Hubble & Hattie is the home of a range of books that cover all-things animal, produced to the same high quality of content and presentation as our motoring books, and offering the same great value for money.

While the publisher and author have designed this book to provide up-to-date information regarding the subject matter covered, readers should be aware that medical information is constantly evolving. The information in this book is not intended as a substitute for veterinary medical advice. Readers should consult their veterinary surgeon for specific instructions on the treatment and care of their dog. The author and Veloce Publishing Ltd shall have neither liability nor responsibility with respect to any loss, damage, or injury caused, or alleged to be caused directly or indirectly by the information contained in this book.

www.hubbleandhattie.com

First published in August 2010 by Veloce Publishing Limited, Veloce House, Parkway Farm Business Park, Middle Farm Way, Poundbury, Dorchester, Dorset, DT1 3AR, England. Fax 01305 250479/e-mail info@veloce.co.uk/web www.veloce.co.uk or www.velocebooks.com. This edition published October 2019.
ISBN: 978-1-787116-01-6 UPC: 6-36847-01601-2

Readers with ideas for books about animals, or animal-related topics, are invited to write to the editorial director of Veloce Publishing at the above address.
British Library Cataloguing in Publication Data – A catalogue record for this book is available from the British Library. Typesetting, design and page make-up all by Veloce Publishing Ltd on Apple Mac. Printed and bound by CPI Group (UK) Ltd, Croydon, CR0 4YY.

The complete dog massage manual

Julia Robertson

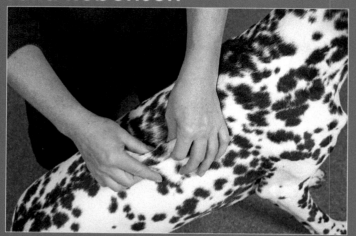
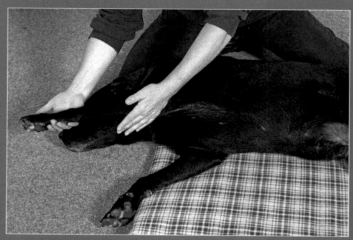
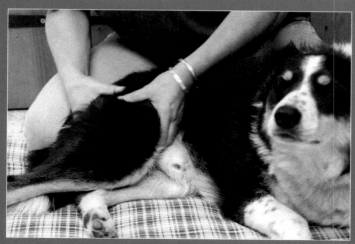
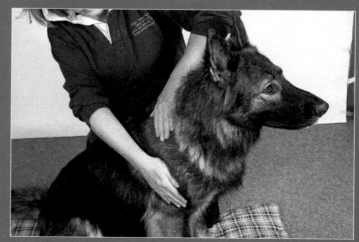

Hubble & Hattie

Gentle Dog Care